INTEGRATING MATH IN THE REAL WORLD

THE MATH OF FOOD

Hope Martin and Susan Guengerich

User's Guide
to
Walch Reproducible Books

As part of our general effort to provide educational materials that are as practical and economical as possible, we have designated this publication a "reproducible book." The designation means that purchase of the book includes purchase of the right to limited reproduction of all pages on which this symbol appears:

Here is the basic Walch policy: We grant to individual purchasers of this book the right to make sufficient copies of reproducible pages for use by all students of a single teacher. This permission is limited to a single teacher and does not apply to entire schools or school systems, so institutions purchasing the book should pass the permission on to a single teacher. Copying of the book or its parts for resale is prohibited.

Any questions regarding this policy or requests to purchase further reproduction rights should be addressed to:

Permissions Editor
J. Weston Walch, Publisher
321 Valley Street • P. O. Box 658
Portland, Maine 04104-0658

1 2 3 4 5 6 7 8 9 10

ISBN 0-8251-3861-2

Printed in the United States of America

Contents

Introduction

In 1989, the National Council of Teachers of Mathematics (NCTM) developed the *Curriculum and Evaluation Standards for School Mathematics* to help teachers take their mathematics classes into the twenty-first century. The document calls for a curriculum that will help students solve problems and make connections between mathematics and other curricular areas, such as science, social studies, language arts, consumer education, and art. For math to be relevant, students must see how it relates to their lives outside of the mathematics classroom.

The lessons, activities, and projects in *Integrating Math in the Real World* have been designed to help students see interrelationships between subjects. Many of the lessons emphasize critical-thinking skills. Many of the activities are open-ended and encourage students to:

- work collaboratively to develop problem-solving strategies
- make connections between their life experiences and the math classroom
- develop self-confidence in their abilities to solve math problems

The Math of Food, one of the books in the *Integrating Math in the Real World* series, provides a comprehensive examination of many different aspects of nutrition, including proper diet, exercise, and information from food labels. *The Math of Food* also examines international issues, such as the problems of world hunger. The material in this book is designed to help students develop good nutritional habits and a regular exercise routine while they are still young enough to develop lifelong habits. The lessons and activities will get students actively involved in significant math projects while providing them with important links to nutritional information.

Each of the lessons is introduced with a Teacher Page. These Teacher Pages include the following sections:

Areas of Study

This section lists the mathematics skills of the lesson. Many of the activities are multifaceted and make use of a variety of math skills.

Concepts

This section contains a concise list of tasks for which the student will be accountable. If a rubric or grading matrix is being used, this list will be invaluable in developing specific criteria for assessment.

Materials

This section contains a list of materials that each student will need for the lesson unless otherwise specified. Collecting these items in advance will assure a smoother flow to the lesson.

Procedures

This section gives a brief description of the lesson with suggestions for the teacher. It is not meant to be a step-by-step recipe but merely a guide to help organize the lesson.

Assessment

Suggestions made are for effective, nontraditional ways to evaluate student achievement. It is suggested that student products be examined and that students be observed during the activity and questioned about their progress. Each lesson has one or more recommendations for journal questions that require a more in-depth understanding of the lesson concepts.

Extensions

Often some students need a more advanced or extended lesson. The suggestions discussed in this section can be used with a select group of students or with the entire class (if the lesson has been motivating and successful).

The lessons in *Integrating Math in the Real World: The Math of Food* have been designed to be teacher- and student-friendly. In many cases, these lessons can be substituted for more traditional lessons found in mathematics texts.

Observation of Students

When students are active and working together, it is essential that the teacher walk around the room to become aware of the progress of the student groups and any problems that might arise. During these times it is possible to assess student understanding in a more formal way. While not every student can be observed each time, it is possible to perform a formal-type assessment at least twice during each grading period for each student. These observations can be shared with both parents and students during parent-teacher conferences.

A form, such as the one below, can be used to make the observations more consistent and simplify the process.

Name of Student ______________________________

Criteria	4	3	2	1
How actively are students participating in group project?				
How well does student appear to understand concept of lesson?				
Is student actively listening to other members of group?				
Is student assuming positive leadership or problem-solving role?				

Comments:

Using a Rubric for Performance Assessment

Authentic assessment is based on the performance of the student and should be closely tied to the objectives of the lesson or activity. A rubric can be used to quantify the quality of the work. If the rubric is explained before the activity or project, the students become aware of the requirements of the lesson. A grading matrix should be developed in which each of the objectives is examined using a five-point scale.

5 Student shows mastery and extends the concepts of the activity in new and unique ways

4 Student shows mastery of the concepts of the lesson

3 Student shows understanding, but there is a flaw in the presentation or reasoning

2 Student shows some understanding and has attempted completion, but there are some serious flaws in the presentation or reasoning

1 Student makes an attempt but exhibits no understanding

0 Student makes no attempt

NCTM Standards Correlation

	Mathematics as Problem Solving	Mathematics as Communication	Mathematics as Reasoning	Mathematical Connections	Numbers and Number Relationships	Number Systems and Number Theory	Computation and Estimation	Patterns and Functions	Algebra/Pre-Algebra	Statistics and Probability	Geometry	Measurement
What Do I Eat?	•	•	•	•	•	•	•			•		
Using the Food Pyramid as a Guide	•	•	•	•			•			•		
What Is a Healthy Diet?	•	•	•	•	•	•	•		•	•		
Gummy Bears	•	•	•	•	•	•	•		•	•	•	•
Using the New Food Label	•	•	•	•	•		•					
Toll House™ Cookie Count	•	•	•	•	•	•	•		•			•
Measurement Dilemma	•	•	•	•	•	•	•	•	•			•
Cost of Cereal	•	•	•	•	•	•	•	•	•	•		•
Cereal Box Study	•	•	•	•	•	•	•		•		•	•
Burn Those Calories	•	•	•	•	•	•	•	•	•	•		
Cardiovascular Fitness	•	•	•	•	•	•	•		•	•		•
Worldwide Nutritional Concerns	•	•	•	•			•			•		
The Distribution of Food	•	•	•	•	•	•	•		•	•		
A Nutrition Word Search				•								
Nutrition Crossword				•								
Nutrition and Poetry				•								
Nutrition and Music				•								
Nutrition and Art				•								
Research Projects	•	•	•	•	•	•	•		•		•	•

What Do I Eat?

Areas of Study

Data collection, reading tables and charts, computation, estimation, fractions, problem solving

Concepts

Students will:

- collect personal nutrition data
- use charts and tables to find information
- calculate the total calories, grams of fat, etc., in their diet for one day
- present an accurate record of one day's diet

Materials

- What Do I Eat? handouts
- additional resource books containing nutritional information
- calculator
- overhead transparencies of the What Do I Eat? handout for the teacher

Procedures

Read through the What Do I Eat? handouts with students. Discuss the problems involved with keeping accurate records over a long period of time and have students make suggestions as to how they might be able to record their intake of food. Stress the importance of accuracy.

While a limited number of foods are supplied on the Nutrition Counter sheets, it will be necessary to have additional resource materials available for students to use. Two recommendations are *The Supermarket Nutrition Counter,* by Natow and Heslin, and the *Encyclopedia of Food Values,* by Netzer. Students can also be encouraged to use the information on the labels of the foods they eat.

This activity serves as an introduction to lessons that follow in which students are asked to design a healthy menu for themselves.

Assessment

1. Student products:
   - completion of accurate data collection tables
2. Observation of students
3. Journal question:
   - Explain the strategies you used to collect your data. Do you believe the food data you collected gives an accurate picture of your normal diet? Why or why not?

Extension

- Students can collect data on their diets for three days and then average the results. While this is more difficult to do, it will probably reflect a more accurate picture of their diets.

Name ______________________ Date ______________

What Do I Eat?

The teenage years are a time of fast growth. What you eat, or how you fuel the engine of your body, can affect how healthy and strong you are. The statement "You are what you eat" should be "What you become depends on what you eat." This activity will help you develop a healthy diet. But before you can figure out what you should be doing, you need to gather some data on what you are doing now (diet-wise, that is).

You must collect some data for this activity. Be sure to keep track of everything you eat for the entire day. You must try to estimate the quantities of food you eat as well. For example, one piece of bread has about 70 calories, but if you have a sandwich, you have to count two pieces of bread (140 calories) and what is between the pieces of bread, also.

The table on the next page will help you organize your data. Be sure to:

1. Write down everything you eat for an entire day.
2. Eat what you would normally eat. Do not change because you are writing down what you are doing.
3. Estimate quantities as best you can.
4. Use the information on the Nutrition Counter or in reference books in your classroom to help find calories, protein, fat, calcium, and carbohydrates in the foods you eat.

After you collect this data, you will learn more about nutritional values and your personal eating habits.

Name ______________________ Date ______________

What Do I Eat?

Foods	Calories	Protein	Fat	Carbohydrates	Calcium
Breakfast					
Lunch					
Snacks					
Dinner					

Name ______________________ Date ______________

What Do I Eat?

The Nutrition Counter

The following table shows a variety of foods, the size of a small portion, and the calories, protein, fat, carbohydrates, and calcium contained in that size portion. You can use this information to help you with many of the activities in this unit. These data are approximate as they may vary with brands.

Food	Portion Size	Calories	Protein g	Fat g	Carbohydrates g	Calcium mg
Apple	1	80	0.3	-	21	10
Bagel	1	195	9	1	38	29
Banana	1	110	1.2	-	-	7
Baked beans	½ cup	160	6.1	2	28	64
Beef patty	4 oz.	312	27.5	22	0	13
Bologna	1 slice	90	3.7	8	1	3
Bread (whole wheat)	1	90	3	2	18	20
Bread (Wonder® white)	1	70	3	1	13	32
Breakfast bars	1	150	2	5	25	20
Broccoli (cooked)	½ cup	22	3	-	4	36
Brownie	1	150	2	7	25	13
Butter	1 pat	36	-	4	-	3
Cake (pound)	1 slice	130	2	7	14	8
Cake (Pop-Tarts®)	1	210	2	6	37	-
Cake (Little Debbie's®)	1 pkg. (2.6 oz)	300	2	15	29	-
Cake (Twinkies®)	1	140	2	4	25	-
Cake (cheesecake)	1/12 cake	456	3	9	32	52
Cantaloupe	½	94	2.3	1	22	28
Carrots	1	30	0.7	-	7	19
Cauliflower (cooked)	½ cup	14	1.5	-	3	14
Cereal (hot instant)	1 pkg.	100	4	2	25	170
Cereal (Alpha-Bits®)	1 cup	110	2	1	25	10

(continued)

Name ______________________________ Date ______________

What Do I Eat?

The Nutrition Counter *(continued)*

Food	Portion Size	Calories	Protein g	Fat g	Carbohydrates g	Calcium mg
Cereal (Raisin Bran®)	3/4 cup	120	3	1	32	16
Cereal (cornflakes)	1 cup	100	2	0	24	3
Cereal (Special K®)	1 cup	100	6	0	20	10
Cheese (American)	.75 oz.	70	6.4	5	-	174
Chicken	3 oz.	135	16	12	-	17
Chocolate chip cookie	1	78	1	5	9	13
Cream cheese	1 oz.	100	2	10	1	23
Eggs (scrambled)	2	200	6.8	15	2	44
Fish sticks	4	200	7	10	18	6
Hot dog (beef)	1	150	6	14	1	24
Ice cream (chocolate)	1/2 cup	140	2	7	19	88
Ice cream (vanilla)	1/2 cup	140	2	7	16	88
Margarine	1 oz.	100	-	11	-	-
Mayonnaise	1 T	100	-	11	-	-
Milk (2%)	1 cup	120	8	5	12	297
Milk (1%)	1 cup	100	8	3	-	300
Milk (whole)	1 cup	150	8	8	11	290
Milk (chocolate)	1 cup	190	8	5	29	-
Orange	1	65	1.2	-	16	56
Pasta	2 oz.	210	4.3	1	-	10
Peanut butter sandwich	1	340	9	19	33	10
Pickle	1/4	4	-	0	1	6
Pizza	1/5	320	13	16	29	200
Popcorn	3 cups	100	2	6	11	1
French fries	1 large	355	4.6	19	44	18
Potato chips	17 chips	150	2	9	15	7
Pudding (chocolate snack pack)	1	170	3	6	28	75
Raisins	1/2 cup	260	3	0	63	36

Using the Food Pyramid as a Guide

Areas of Study

Analysis of data, open-ended problem solving, decision making, reading charts and tables

Concepts

Students will:

- compare a sample menu with the Food Guide Pyramid
- classify the sample menu by the categories on the Food Pyramid
- compare two menus and suggest improvements
- respond to the Food Pyramid Journal

Materials

- Food Pyramid handouts
- Using the Food Pyramid as a Guide handouts
- additional reference books for research

Procedures

Discuss the food pyramid with the class. Give students a copy of the sample menu with check marks indicating the food group categories to which foods belong. Work with them to total the servings in each group and compare them with the recommendations on the Food Pyramid.

The sample shown on page 8 does not meet the guidelines of the Food Pyramid. It is:

- short on bread, cereal, rice, pasta
- short on vegetables
- over on fats

Once students understand the process, they can analyze the second menu, placing foods in the correct food group(s). Emphasize that foods may fall into more than one category and a menu may contain more than one serving in a category. For example, a peanut butter and jelly sandwich has two servings from the bread category, one serving from the meat, fish, poultry, and nuts category, and two servings from the fats/sweets category.

The Food Pyramid Journal is a more open-ended activity that asks students to compare the two menus (pages 8 and 9) and make suggestions for improvements. In this way, students can see how slight changes in their menu can result in a healthier diet. Also, they are asked to include their preferences and how these might change the nutritional aspects of the menu.

Assessment

1. Student products
2. Observation of students
3. Journal questions: included in the lesson

Extension

- Students can analyze their diet from the activity What Do I Eat? They would need to design a table for their analysis.

Name ______________________ Date ______________

Using the Food Pyramid as a Guide

The Food Pyramid

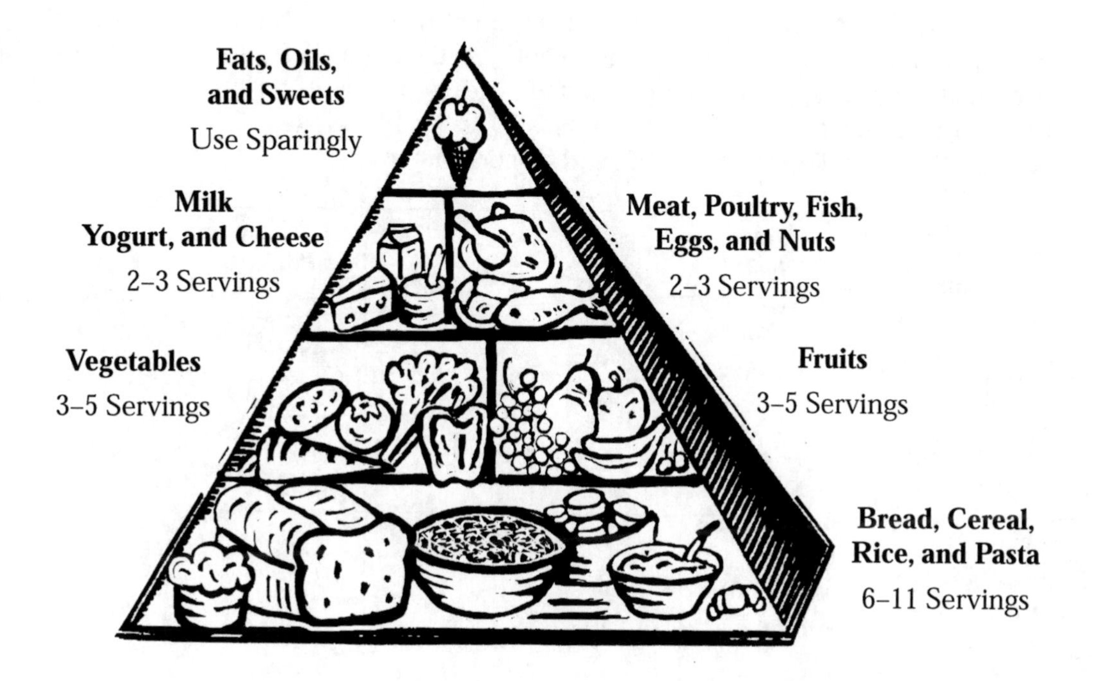

As you plan your ideal menu, keep this Food Guide Pyramid in mind. Everything on it is necessary for good nutrition. Use it as a general guide, but keep in mind the other nutritional recommendations.

Name ______________________________ Date ______________

Using the Food Pyramid as a Guide

Below is a sample menu for one day. This chart lists the categories on the food pyramid. Each food has been placed in the correct category. Some foods fall into more than one category. For example, a turkey sandwich is made of bread, turkey, and mayonnaise. This means that the turkey sandwich has two checks in the Bread, Cereal, Rice, and Pasta category (because it is made with **two** slices of bread), one check in the Meat, Poultry, Fish, Eggs, and Nuts category, and one check in the Fats, Oils, and Sweets category.

Foods	Bread, Cereal, Rice, and Pasta	Fruits	Vegetables	Meat, Poultry, Fish, Eggs, and Nuts	Milk, Yogurt, and Cheese	Fats, Oils, and Sweets
Breakfast						
Orange juice		✔				
Cereal with milk	✔				✔	
Banana		✔				
Lunch						
Turkey sandwich with mayonnaise	✔✔			✔		✔
Carrot sticks			✔			
Apple		✔				
Candy bar						✔
Milk					✔	
Snack						
Pretzels	✔					
Orange		✔				
Dinner						
Baked chicken				✔		
Rice	✔					
Green beans			✔			
Ice cream					✔	✔
TOTAL SERVINGS						

Does this menu meet the recommended servings on the Food Guide Pyramid in each category? What could be added or eliminated to fit the requirements of the Food Guide Pyramid? (Use the back of this sheet for your answer.)

Name ____________________ Date ____________

Using the Food Pyramid as a Guide

Now it's your turn to do the classifying. Below is another sample menu. Think about each item on the menu carefully and check off where it belongs on the Food Pyramid. Be sure to check each category the food falls into, as well as whether it is more than one serving. For example, a sandwich has **two** pieces of bread and therefore would be **two servings** in the Bread, Cereal, Rice, and Pasta group.

Foods	Bread, Cereal, Rice, and Pasta	Fruits	Vegetables	Meat, Poultry, Fish, Eggs, and Nuts	Milk, Yogurt, and Cheese	Fats, Oils, and Sweets
Breakfast Orange juice Toast with butter and jelly Milk						
Lunch Peanut butter and jelly sandwich Potato chips Apple Brownie Milk						
Snack Buttered popcorn Yogurt						
Dinner Hamburger on a bun Baked potato Green salad with tomato and 1 tablespoon dressing Soda						
TOTAL SERVINGS						

Name ______________________ Date ______________

Using the Food Pyramid as a Guide

My Food Pyramid Journal

Examine the two sample menus and then use the space below to discuss their strengths and/or weaknesses. How might you change the menu to include foods you like and would normally eat? Are your food choices better or worse than the samples? Use the rest of the space on this page to say why or why not.

What Is a Healthy Diet?

Areas of Study

Collection and organization and analysis of data, open-ended problem solving, decision making, percent of difference, computation, pre-algebra concepts

Concepts

Students will:

- determine the number of calories needed to maintain a desired body weight
- compare their diets to nutritional guidelines and calculate the percent of difference
- design a healthy diet using nutritional guidelines and the food pyramid
- plan a one-day menu that meets all the requirements for a healthy diet

Materials

- What Is a Healthy Diet? handouts
- calculator
- nutritional reference books for research

Procedures

Students may be aware of adult requirements in terms of calories and ideal weight. However, growing youngsters need many more calories than adults, and their misconceptions may result in diets which lack necessary nutrients and calories.

Keep in mind that setting up nutritionally correct one-day menus is a difficult job. Students must keep track of calories, grams of protein, carbohydrates, and fat. They must keep the levels of sodium down and find a menu that is reasonably acceptable to a teenager.

Review the table of nutritional requirements for 11- to 14-year-olds.

Note: The total calories are the numbers required for maintenance of body weight.

- Students find the number of calories they need each day to maintain their desired body weights by multiplying their desired body weight by 25 (boys) or 22 (girls).
- Students find the number of grams of carbohydrates they need daily by multiplying the number of calories they found above by 0.55 (55%) and dividing that number by 4 because there are four calories in each gram of carbohydrate.
- Students find the number of grams of protein they need daily by multiplying the number of calories they found by 0.15 (15%) and dividing that number by 4.
- Students find the number of grams of fat they need each day by multiplying the number of calories by 0.30 (30%) and dividing that number by 9 because there are nine calories in each gram of fat.

The My Personal Menu for One Day activity requires students to plan a menu using nutritional information available on charts, in books, and on food labels. This is not a trivial task. Some flexibility should be permitted regarding portion size and exact quantities.

This activity is designed to have students examine and analyze their diets and make adjustments to bring them into closer alignment with accepted nutritional guidelines.

Assessment

1. Student products:
   - completed worksheets, including Personal Menu
2. Observation of students

3. Journal questions:
   (a) If a patient wanted to lose weight, why would a doctor recommend limiting the amount of fat in his or her diet and increasing the amount of carbohydrates?
   (b) Is the daily menu you developed a diet you could live with? Why or why not?

Extensions

- Have students develop a menu for one week.
- Have students pretend they are allergic to a particular food, such as wheat flour or dairy products, and have them develop a daily menu that meets nutritional guidelines without that food.
- Have students research cholesterol and design a menu that is very low in cholesterol.

Name ______________________ Date ______________

What Is a Healthy Diet?

- Is your diet healthy? What do the experts consider a healthy diet? The information below has been supplied by the Food and Nutrition Board of the National Research Council and the *Encyclopedia of Food Values,* by Corinne T. Netzer.
- Important: Some nutritionists recommend taking your desired body weight and multiplying it by 15 to determine the number of calories you need to maintain this weight. This guideline is meant to be used by adults who have completed their growth cycle.

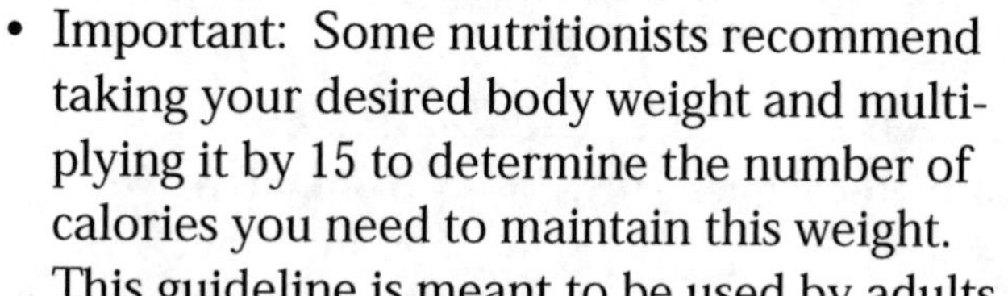

- At your age, you need either 25 or 22 calories for each pound of body weight. This is because you are growing so quickly. So, if you are a 100-pound girl, you need 2,200 calories each day.
- The table below shows how many grams or milligrams of protein, calcium, carbohydrates, and fat are required for boys and girls between 11 and 14 years old.

	11–14 Years Old	
Component	**Boys**	**Girls**
Total calories	25 per pound of body weight	22 per pound of body weight
Protein (g)	45	46
Calcium (mg)	1,200	1,200
Carbohydrates (g)	375	330
Fat (g)	less than 80	less than 75

Name ______________________ Date ______________

What Is a Healthy Diet?

- How does your diet compare with what you have learned? Let's compare your nutritional requirements with the data you collected when you kept a record of your diet for one day.
- To find the percent of the difference:
- $\% \text{ of difference} = \dfrac{\text{difference between actual and recommended number}}{\text{recommended number}}$

What I Ate	What Is Recommended	% of Difference
Number of calories I ate ______	Calorie requirements ______	______
My protein intake ______	Protein requirements ______	______
My calcium intake ______	Calcium requirements ______	______
My carbohydrate intake ______	Carbohydrate requirements ______	______
My fat intake ______	Fat requirements ______	______

- Can you use this information to help you design a one-day menu that would be considered healthy by nutritionists? You can use any reference material you have at your disposal. Try to choose a menu of foods that you like.
- Use the information on Designing a Healthy Diet—Just for Me to design your menu. Fill in your eating plan on My Personal Menu for One Day.

Name ______________________________ Date ______________

What Is a Healthy Diet?

Designing a Healthy Diet—Just for Me!

When you design your healthy diet, you want to plan a diet of foods that you like. You also want to make sure that you meet the dietary requirements to help you grow into a healthy adult. As you plan your menu, follow these guidelines:

- What should you eat?
  1. 55% of your total calories should come from carbohydrates (the foods on the bottom level of the food pyramid).
  2. 15% of your total calories should come from protein.
  3. A maximum of 30% of your total calories can come from fats.
  4. Fruits and vegetables are so good for you, they do not have to be calculated in this way. But do not forget to include at least 6 servings of fruits and vegetables as part of your menu or you will lose out on needed vitamins and minerals.
- How much should you eat?
  1. If you are an active girl, you should eat 22 calories for each pound of desired weight.
  2. If you are an active boy, you should eat 25 calories for each pound of desired weight.
- What other things do we need to consider?
  1. Watch the sodium—no more than between 1,100 and 3,300 mg per day. Remember that each teaspoon of salt contains about 2,100 mg of sodium.
  2. Dietary fiber helps prevent certain types of cancer and helps our bodies digest foods. Try to eat between 25 and 35 grams of fiber each day.
  3. A growing person needs good blood and strong bones and teeth—that's the role of calcium. Try to eat foods high in calcium, such as citrus fruits, tomatoes, leafy green vegetables, broccoli, and potatoes.

Name ______________________________ Date ______________

What Is a Healthy Diet?

Designing a Healthy Diet—Just for Me! *(continued)*

- How do we get started?

  Let's use an example of an active girl who wants to weigh 100 pounds.

- She will need to eat 2,200 calories each day to maintain her weight (22×100).

  55% of the total calories should come from carbohydrates.

  $2{,}200 \times 0.55 = 1{,}210$ calories; $1{,}210 \div 4 \approx 300$ grams. We divide by 4 because each gram of carbohydrate has four calories.

- 15% of the total calories should come from protein.

  $2{,}200 \times 0.15 = 330$ calories; $330 \div 4 \approx 83$ grams. Again we divide by 4 because each gram of protein has four calories.

- 30% of the total calories may come from fat.

  $2{,}200 \times 0.30 = 660$ calories; $660 \div 9 \approx 73$ grams. This time we divide by 9 because each gram of fat has nine calories.

- Keep track of the sodium and calcium as well.

  Remember to use the Food Guide Pyramid and you will probably have all the other necessary vitamins and minerals you need.

  So, our girl should eat 2,200 calories of food: 300 grams of carbohydrates, 83 grams of protein, and up to 73 grams of fat.

Name ______________________ Date ______________

What Is a Healthy Diet?

Designing a Healthy Diet—Just for Me! *(continued)*

Use the space below to calculate your own personal healthy diet. Write your "perfect quantities" in the blanks at the bottom of the page.

My Calculations

(Use this space to show your work.)

1. How many calories do you need?
2. 55% of your calories should come from carbohydrates. How many calories should come from carbohydrates?

   Now divide this number by 4 (there are four calories in each gram of carbohydrates) to find the number of grams of carbohydrates. The number of grams of carbohydrates should be:
3. 15% of your calories should come from protein. How many calories should come from protein?

   Now divide this number by 4 (there are four calories in each gram of protein) to find the number of grams of protein. The number of grams of protein should be:
4. 30% of your calories may come from fats. How many calories may come from fats?

   Now divide this number by 9 (there are nine calories in each gram of fat) to find the number of grams of fat. The number of grams of fat may be:
5. Watch the sodium—no more than 3,300 milligrams per day. How much sodium is there in this diet?
6. Dietary fiber is important—between 25 and 35 grams each day. How much dietary fiber is there in this diet?

My ideal weight is ______________________ pounds.

The number of calories I need to eat is ______________________.

The number of grams of protein is ______________________.

The number of grams of fat is ______________________.

The number of grams of carbohydrates is ______________________.

Name ______________________________ Date ________________

What Is a Healthy Diet?

My Personal Menu for One Day

Food	Portion size	Calories	Protein g	Fat g	Carbohydrates g
Breakfast					
Lunch					
Dinner					
Snacks					
Totals					

Gummy Bears

Areas of Study

Data collection, fractions, decimals, percents, graphing, measurement, connections with nutrition

Concepts

Students will:

- collect and analyze data
- convert data to fractions, decimals, and percents
- draw a pie graph
- analyze the contents of a package for weight
- analyze the nutritional information given on a food label

Materials

- Gummy Bears handouts
- bag of 20 gummy bears for each pair of students
- calculator
- markers, crayons, or colored pencils

Procedures

Before beginning the lesson, make up one package of 20 gummy bears for each pair of students.

Have students work in pairs to predict the number of each flavor of gummy bears in their package. They should record their estimate on the data collection page before they begin the experiment. Once the prediction is made, students can open their packages and begin the activity.

Each pair of students should be given a calculator to help with the computation. Since each package has 20 gummy bears, the calculations on the worksheets should not be a problem.

After completing the table (page 20), students fill in the pie graph (page 21) to represent their data.

Have students answer the questions based on the nutrition label (page 22). Follow-up discussion can include the mass of each bear, whether the snack is healthy, etc.

Assessment

1. Student product:
   - successful completion of data collection sheet and graph
2. Observation of students
3. Journal question:

   Your package of gummy bears contained 20 bears and had 173 calories. It weighed 55 g.

   (a) How many bears would you estimate you would count in a kg bag?

   (b) How many calories might you consume if you ate the whole kg bag?

Extension

- Students can research other nonfat snacks that they believe would be healthier than this one.

Name ____________________ Date ____________________

Gummy Bears

Before opening your package, guess the number of gummy bears inside. Next, open the package and count the bears. Then, fill in the data collection table.

Estimated number of gummy bears: ____________________

Flavor of Gummy Bear	Number	Fraction	Decimal	Percent	Degrees (°) of Circle
Cherry					
Lemon					
Grape					
Orange					
Strawberry					
Lime					
TOTAL					

Name ______________________________ Date ______________

Gummy Bears

Use the information from your data collection table to fill in this pie chart.

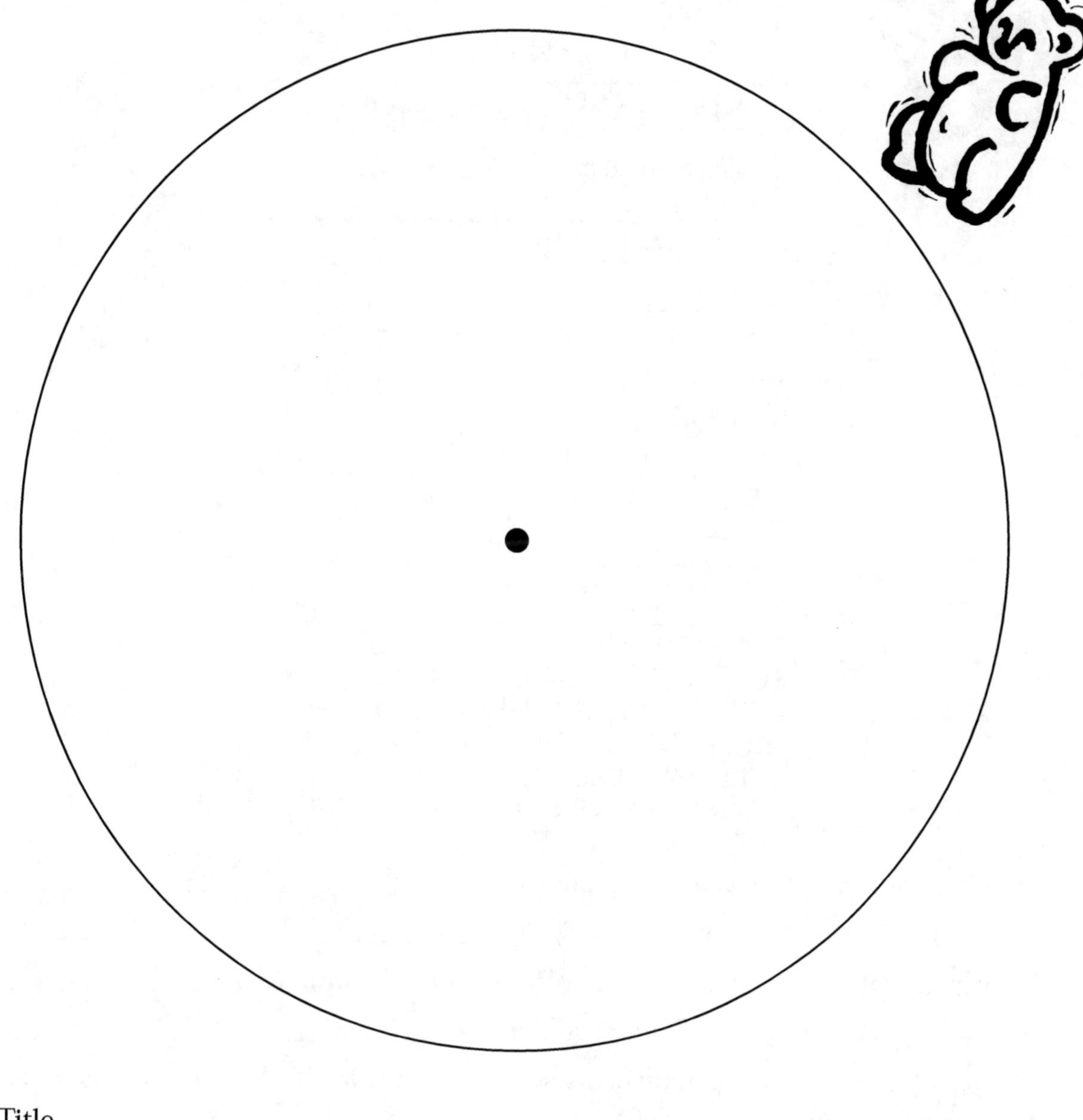

Title ______________________________

The Artist(s) ______________________________

Name ______________________ Date ______________

Gummy Bears

Use the information on the nutrition label to help you answer the questions below.

Nutrition Label

Nutrition Facts	
Serving Size 20 pieces (55g) Servings per Container 1	
Amount per Serving	
Calories 173	
	% Daily Values
Total Fat 0g	**0%**
Sodium 7mg	**0%**
Total Carbohydrates 41g	**10%**
Sugars 29g	
Protein 3g	
Percent Daily Values are based on a 2,000 calorie diet.	
INGREDIENTS: CORN SYRUP, SUGAR, GELATIN, SORBITOL, CITRIC ACID, NATURAL AND ARTIFICIAL FLAVORS, RED 40, YELLOW 5, BLUE 1, CONFECTIONER'S GLAZE.	

How much does your package of gummy bears weigh? ______________

How much does each gummy bear weigh? ______________

How many calories did your package of gummy bears contain? ______________

How many calories are there in one gummy bear? ______________

The package claims that gummy bears are "a great tasting healthy snack." Do you think this is true? Why or why not? ______________

____

____

Using the New Food Label

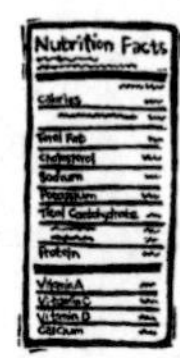

Areas of Study

Reading tables and charts, computation, percentages, problem solving

Concepts

Students will:

- analyze the nutritional information on a food label
- calculate the percent of calories from fat, carbohydrates, and protein
- compare recommended amounts of sodium to the amount of sodium noted on the label
- analyze their own food labels

Materials

- Using the New Food Label handouts
- calculator
- nutrition labels cut from packages
- overhead transparency of food label for teacher

Procedures

Read through Using the New Food Label with students and study the information provided. A worksheet containing nutritional information about Swanson's Homestyle Turkey Dinner™ is provided. This dinner has 290 calories and 11 grams of fat. Since each gram of fat contains 9 calories, 99 of the calories are from fat (9×11).

We can set up the following ratio:

$$\% \text{ of fat} = \frac{\text{calories from fat}}{\text{total calories}}$$

$$\% \text{ of fat} = \frac{99}{290} \approx 34\%$$

The students are asked, "Would you consider this a low-fat meal?" They must reflect back on nutritional guidelines which recommend 30% or less fat to make a knowledgeable response.

A page is provided on which students can paste their own labels and analyze the nutritional value of a food of their choice.

Solutions, p. 25

- 99 calories, 34%, no
- 120 calories, ≈41%
- the fat; each gram of fat contains 9 calories
- between 1,100 and 3,300 mg/day, between 31% and 92%

Assessment

1. Student products:
   - completed worksheet
   - analysis of student's own label
2. Observation of students
3. Journal questions:
   (a) Do you consider the food you picked healthy? Why or why not?
   (b) Do you think that the serving size listed on the label is a normal-size portion? Why or why not?

Extensions

- Bring in the labels from several varieties of vegetables and compare their nutritional values.
- Bring in the labels of foods described as light or reduced fat. How do these compare with the fat in the regular product? For example, how does low-fat mayonnaise compare in fat content to regular mayonnaise?

Name ______________________ Date ______________

Using the New Food Label

Figuring out which foods can be part of a healthy diet can be confusing. In the past, the information on food labels did not make the job of choosing healthy foods any easier. In 1994, however, food labels became easier to read. The following is an example of the new food label.

Serving sizes are now stated in both household and metric measures, and they reflect the amounts people actually eat.

The list of nutrients covers those most important to the health of today's consumers. Most consumers today need to worry about getting too much of certain nutrients (fat, for example), rather than too few vitamins or minerals, as in the past.

The label of larger packages may now tell the number of calories per gram of fat, carbohydrate, and protein.

Nutrition Facts

Serving Size 1 cup (228g)
Servings per Container 2

Amount per Serving	
Calories 260 Calories from Fat 120	
	% Daily Value*
Total Fat 13g	**20%**
Saturated Fat 5g	**25%**
Cholesterol 30mg	**10%**
Sodium 660mg	**28%**
Total Carbohydrate 31g	**10%**
Dietary Fiber 0g	**0%**
Sugars 5g	
Protein 5g	

Vitamin A 4% • Vitamin C 2%
Calcium 15% • Iron 4%

* Percent Daily Values are based on a 2,000 calorie diet. Your daily values may be higher or lower depending on your calorie needs:

	Calories	2,000	2,500
Total Fat	Less than	65g	80g
Sat Fat	Less than	20g	25g
Cholesterol	Less than	300mg	300mg
Sodium	Less than	2,400mg	2,400mg
Total Carbohydrate		300g	375g
Dietary Fiber		25g	30g

Calories per Gram:
Fat 9 • Carbohydrate 4 • Protein 4

• This label is only a sample.
Source: Food and Drug Administration, 1994

Calories from fat are now shown on the label to help consumers meet dietary guidelines. These guidelines recommend that people get no more than 30 percent of the calories in their overall diet from fat.

% Daily Value shows how a food fits into the overall daily diet.

Daily Values are also something new. Some are maximums, as with fat (65 grams or less); others are minimums, as with carbohydrates (300 grams or more). The daily values for a 2,000- and a 2,500-calorie diet must be listed on the label of larger packages.

Name ______________________________ Date ______________

Using the New Food Label

A box of Swanson's Homestyle Turkey Dinner™ with stuffing and potatoes supplies the following nutritional information:

Calories	Protein	Carbohydrates	Fat	Cholesterol	Sodium	Fiber
290	18g	30g	11g	unavailable	1,010mg	unavailable

- This frozen dinner has 11 grams of fat. There are 9 calories for each gram of fat.
  How many calories come from fat? ______________________
  What percentage of the total calories is from fat? ______________________
  Would you consider this a low-fat meal? Why or why not? ______________
  ____
  ____

- This frozen dinner has 30 grams of carbohydrates. There are 4 calories for each gram of carbohydrates.
  How many calories in this frozen dinner come from carbohydrates? ________
  What percentage of the total calories are from carbohydrates? ____________

- This dinner has more grams of protein than grams of fat. Which one accounts for a greater percentage of the total calories? ______________________
  Why? ____
  ____

- This frozen dinner contains 1,010 milligrams of sodium.
  About how many milligrams of sodium do nutritionists say we should have per day? ____
  About what percent of the daily allowance of sodium does this meal have?
  ____
  Do you think this is a reasonable amount? Why or why not? ______________
  ____
  ____

Name ______________________ Date ______________

Using the New Food Label

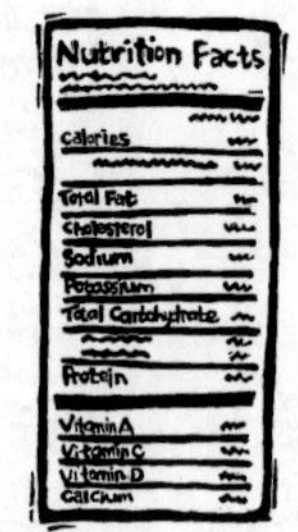

Find your own label and analyze the nutritional value of the product. Paste your nutrition label in the space at the left of this page. Then use the information on the label to answer the questions below.

My Food Label

My food is ______________________

Total calories	
Grams of protein	
Grams of carbohydrates	
Grams of fat	
Milligrams of cholesterol	
Milligrams of sodium	
Grams of fiber	

What percentage of total calories are from

- fat? ______________________
- carbohydrates? ______________________
- protein? ______________________

Analyze the nutritional value of your food. Be sure to discuss whether it meets nutritional guidelines. (Use the back of this sheet.)

Toll House™ Cookie Count

Areas of Study

Fractions, conversions, reading tables and charts, problem solving, computation

Concepts

Students will:

- calculate the calories, fat, sodium, and carbohydrates in one Toll House cookie
- convert the data from standard measurements to fit the quantities in the recipe
- read and interpret data from a supplied table

Materials

- Toll House Cookie Count handouts
- calculator
- additional recipes if students are to complete extension activity

Procedures

Look over the recipe in this activity and discuss with students what steps will be needed to calculate the nutritional data in one Toll House cookie. Go over the conversion information, making sure that students understand how they would find the number of calories in 1 cup of margarine, given that 1 tablespoon has 100 calories. Since there are 16 tablespoons in 1 cup, there would be 1,600 calories in 1 cup of margarine.

Once students have completed the exercise, discuss whether a person would eat just one of these cookies. Do they think these are very large or rather small cookies? What might be a normal serving of this type of cookie? How many calories, grams of fat, milligrams of sodium, and grams of carbohydrates would be consumed in what they consider a normal portion of cookies?

Assessment

1. Student products:
   - completion of worksheets
2. Observation of students
3. Journal questions:
   (a) Why do you think recipe books make portion sizes so small when they provide nutritional information?
   (b) Why should people look at more than just fat content when they choose their foods?

Solutions

Ingredient	Amount	Calories	Fat (g)	Sodium (mg)	Carbohydrates (g)
Flour	2 1/4 cups	900	2.25	0	195.75
Baking soda	1 teaspoon	0	0	435	1
Salt	1 teaspoon	0	0	2,132	0
Margarine	1 cup	1,600	176	1,520	0
White sugar	3/4 cup	577.5	0	3.75	149.25
Brown sugar	3/4 cup	615	0	72.75	159
Vanilla	1 teaspoon	10	0	0	0
Water	1/2 teaspoon	0	0	0	0
Eggs	2	150	10	126	2
Chocolate chips	12 ounces	1,716	60	300	216
Total		5,568.5	248.25	4,589.5	723

One Toll House cookie

Calories	Fat (g)	Sodium (mg)	Carbohydrates (g)
about 56 cal.	about 3 g	about 46 mg	about 7 g

Name ______________________________ Date ______________________________

Toll House Cookie Count

You have just been hired as a reporter for the local paper. Your first assignment is to figure out the nutritional value of the ingredients in the Toll House cookie recipe below. This recipe makes 100 cookies. To complete your assignment, you will need the conversion information at the bottom of this page and the nutritional data on the next page.

Ingredient	Amount	Calories	Fat (g)	Sodium (mg)	Carbohydrates (g)
Flour	2 ¼ cups				
Baking soda	1 teaspoon				
Salt	1 teaspoon				
Margarine	1 cup				
White sugar	¾ cup				
Brown sugar	¾ cup				
Vanilla	1 teaspoon				
Water	½ teaspoon				
Eggs	2				
Chocolate chips	12 ounces				
Total					

You will need to use some of these conversions to help with your calculations:

DRY

3 teaspoons	=	1 tablespoon
4 tablespoons	=	¼ cup
16 tablespoons	=	1 cup

LIQUID

2 tablespoons	=	1 ounce
2 ounces	=	¼ cup
8 ounces	=	1 cup

Name ______________________________ Date ______________

Toll House Cookie Count

The information provided below is from *The Supermarket Nutrition Counter,* by Natow and Heslin. Use it to calculate the nutritional information for each ingredient in the Toll House cookie recipe. The table tells you that 1 cup of flour contains 400 calories, but the recipe calls for $2\frac{1}{4}$ cups of flour. Be sure to calculate totals based on quantities in the recipe.

Nutritional Table					
Ingredient	**Amount**	**Calories**	**Fat (g)**	**Sodium (mg)**	**Carbohydrates (g)**
Flour	1 cup	400	1	0	87
Baking soda	1 teaspoon	0	0	435	1
Salt	1 teaspoon	0	0	2,132	0
Margarine	1 tablespoon	100	11	95	0
White sugar	1 cup	770	0	5	199
Brown sugar	1 cup	820	0	97	212
Vanilla	1 teaspoon	10	0	0	0
Water	$\frac{1}{2}$ teaspoon	0	0	0	0
Eggs	1	75	5	63	1
Chocolate chips	1 ounce	143	8	25	18

Use this information to learn all about one Toll House cookie!

One Toll House cookie

Calories	Fat (g)	Sodium (mg)	Carbohydrates (g)

Measurement Dilemma

Areas of Study

Fractions, equivalent fractions, and computation

Concepts

Students will:

- convert recipes so they can be measured using only one measuring spoon and one measuring cup
- calculate correct measurements using fraction equivalents

Materials

- Measurement Dilemma handouts
- calculators
- set of measuring spoons and measuring cups
- cookbooks

Procedures

Review the use of equivalent fractions and the converting of measures with different units. Display a set of measuring spoons and measuring cups and explain their use in cooking. Emphasize that only one spoon and one cup measure can be used to make these recipes. Ask students to find a family recipe and convert it also.

Possible Solutions

- Whole Wheat Pancakes (p. 31): $^1/_4$ teaspoon and $^1/_2$ cup
- Ham-Roni Salad (p. 32): $^1/_2$ teaspoon and $^1/_4$ cup
- Eggplant Patties (p. 32): $^1/_2$ teaspoon and $^1/_4$ cup
- Fiesta Corn (p. 33): $^1/_2$ teaspoon and $^1/_2$ cup

Assessment

1. Student product:
   • completed handouts
2. Observation of students
3. Journal questions:
   (a) Is it more work to convert a recipe or to wash many measuring cups?
   (b) Which of the recipes would you like to try? Why?

Extensions

- Convert the recipes so they can be measured using only one measuring spoon.
- Research the origins of cup, quart, teaspoon, and tablespoon measures.

Name ______________________________ Date ______________

Measurement Dilemma

Measuring spoons come in sets of $\frac{1}{4}$ teaspoon (t), $\frac{1}{2}$ teaspoon, 1 teaspoon, and 1 tablespoon (T). Measuring cups come in sets of $\frac{1}{4}$ cup, $\frac{1}{3}$ cup, $\frac{1}{2}$ cup, $\frac{2}{3}$ cup, $\frac{3}{4}$ cup, and 1 cup. Convert the recipe below so that you need to use only one spoon measure and one cup measure. Use the conversion chart to help you.

Conversion Chart

1 tablespoon	= 3 teaspoons
16 tablespoons	= 1 cup
4 cups	= 1 quart

Whole Wheat Pancakes

Measuring spoon selected ____________ Measuring cup selected ____________

Ingredient	Recipe	Your Measurement
Whole wheat flour	$1\frac{1}{2}$ cups	
Baking powder	$\frac{1}{2}$ T	
Baking soda	1 t	
Brown sugar	3 T	
Salt	$\frac{3}{4}$ t	
Buttermilk	$1\frac{1}{2}$ cup	
Vegetable oil	3 T	
Beaten eggs	2	

Mix dry ingredients. Combine eggs, milk, and oil; add to dry ingredients. Mix just until smooth. Fry on lightly greased griddle.

Name ______________________________ Date ______________

Measurement Dilemma

Ham-Roni Salad

Measuring spoon selected __________ Measuring cup selected __________

Ingredient	Recipe	Your Measurement
Diced cooked ham	1 cup	
Cooked macaroni, rinsed in cold water	2 cups	
Diced celery	$\frac{3}{4}$ cup	
Sliced carrots	1 cup	
Chopped green pepper	$\frac{1}{4}$ cup	
Chopped green onions	$\frac{1}{4}$ cup	
Salad dressing or mayonnaise	3 T	
Barbecue sauce	$1\frac{1}{2}$ T	
Prepared mustard	1 t	

Combine ham, macaroni, celery, carrots, green pepper, and onion in salad bowl. In small bowl, mix salad dressing, barbecue sauce, and mustard. Combine all ingredients and chill.

Eggplant Patties

Measuring spoon selected __________ Measuring cup selected __________

Ingredient	Recipe	Your Measurement
Medium eggplant, pared and cubed	1	
Crushed club crackers	$1\frac{1}{4}$ cups	
Shredded cheddar cheese	$1\frac{1}{4}$ cups	
Eggs, slightly beaten	2	
Minced parsley	2 T	
Chopped onion	2 T	
Clove garlic, minced	1	
Salt	$\frac{1}{2}$ t	
Pepper	dash	

In covered pan, cook eggplant in boiling water for 5 minutes. Drain very well and mash. Stir in other ingredients. Shape into patties and fry about 3 minutes on each side. Makes 8 to 10 patties.

Name ________________________________ Date ________________

Measurement Dilemma

Fiesta Corn

Measuring spoon selected ____________ Measuring cup selected ____________

Ingredient	Recipe	Your Measurement
Chopped onion	½ cup	
Margarine or butter	2 T	
Flour	2 T	
Processed American cheese, cut in cubes	1 cup	
Fresh tomatoes, diced	2 cups	
Drained 16-ounce cans whole-kernel corn	2	
Salt	½ t	
Pepper	dash	

Cook onion in margarine. Blend in flour. Stir in cheese cubes and diced tomatoes. Cook until cheese melts. Add corn and seasonings. Continue cooking for 5 more minutes, stirring occasionally.

Your Family Recipe

Find a favorite family recipe and copy it down. Then convert the recipe so that you can make it using only one measuring spoon and one measuring cup.

Name of recipe ________________________________

Measuring spoon selected ____________ Measuring cup selected ____________

Ingredient	Recipe	Your Measurement

Cooking directions: (Use back of this sheet.)

Cost of Cereal

Areas of Study

Computation, data collection, spreadsheets, rounding to the nearest tenth of a cent, mean, median, mode, and range

Concepts

Students will:

- collect information from nutrition tables on cereal boxes
- calculate the cost per serving and cost per ounce of various cereals
- calculate the mean, mode, median, and range of the cost per serving and per ounce for the cereals

Materials

- Cost of Cereal handouts
- calculator
- boxes of cereal with exact prices for class

Procedures

Several weeks before the start of this project, send a letter home asking parents to send in an empty box of cereal with the exact price of the cereal written on the box (page 35). Collect extra boxes with the exact prices for those who have forgotten or could not bring an empty box of cereal. Using the nutrition label on the cereal box, find the number of servings per box and the number of ounces in the box. Each student will then calculate the cost per serving and cost per ounce for the box of cereal he or she brought in. Collect class data. You may find it helpful to review the concepts of mean, median, mode, and range before distributing the data collection sheet. Then, help students find the range, mean, mode, and median of each set of data.

Assessment

1. Student product:
   - completed handouts
2. Observation of students
2. Journal questions:
   (a) Do there seem to be any patterns in the cost per serving data?
   (b) Do the largest boxes seem to have the lowest price per ounce? Can you find an exception to this rule?
   (c) What kinds of cereals tend to have the lowest cost per serving? What kinds of cereals tend to have the highest cost?
   (d) Compare the nutritional value of the highest and lowest cost per serving cereal.

Extensions

- Have students find two comparable cereals, one a national brand with extensive advertising and another a store brand, and compare the price, taste, nutritional value, and packaging. Students can then report their findings to the class.
- Use a computer spreadsheet program to display and calculate the data from this activity.

Name ____ Date ______________________

Cost of Cereal

Dear Parent(s),

Our class will be studying the mathematics of nutrition. Each student will need to bring in an empty box of breakfast cereal. The exact cost of the cereal should be written on the box. The box may be brought to school any time before _________ ____________________________, when this project will begin. The box should be marked with your child's name. If your family does not eat breakfast cereal, please ask a neighbor or a friend to save a cereal box for you.

Thank you for your cooperation with our project. Please call the school if you have any questions.

Sincerely,

Name ______________________________ Date ____________________

Cost of Cereal

Each class member should collect the data from the boxes of cereal brought to school. The cereal brand, weight of the box, cost per box, and servings per box should be noted. Use this spreadsheet to calculate each cereal's cost per serving and cost per ounce. Round the cost per serving and cost per ounce to the nearest tenth of a cent.

Data Collection Sheet

Brand of Cereal	Cost of Box	Servings per Box	Ounces per Box	Cost per Serving	Cost per Ounce

Name ______________________________ Date ______________

Cost of Cereal

Use the data collected from the cereal boxes to find the range, mean, mode, and median for both the cost per ounce and the cost per serving.

Cost per Serving	Cost per Ounce
Range	Range
Mean	Mean
Mode	Mode
Median	Median

Explain the differences in cost per ounce and cost per serving. ______________

____

How does the cost per serving of your cereal compare to the mean cost per serving?

____

One of your classmates brings in a box of Zingo, a new high-energy, highly advertised cereal. This cereal has just been introduced to the public. It costs 92 cents per serving. How does this cost compare to the cost of the cereal you brought to class? How would it affect the range, mean, mode, or median of your class data?

____

Who could use the data your class has collected and how? ______________

____

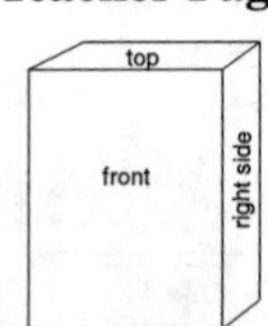

Cereal Box Study

Areas of Study

Measurement, rounding to the nearest centimeter, surface area, volume, reading tables, measurement conversions, computation, percents

Concepts

Students will:

- measure a cereal box and calculate the volume and surface area of the box
- calculate the amount of cereal in the box by reading the nutritional label
- calculate the percent of the box that is full and the percent that is empty

Materials

- Cereal Box Study handout
- calculator
- centimeter ruler
- one cereal box per group

Procedures

Divide class into groups of three or four. Distribute rulers, calculators, and copies of the handout to each group. Review the concepts of surface area, volume, and measuring to the nearest centimeter. Each group measures the dimensions of their box to the nearest centimeter and calculates the volume and surface area of the box. Using the data on the nutrition label, students then calculate the number of cups of cereal in the box. This value is changed to cubic centimeters using the following conversion:

1 cup = 240 cubic centimeters

The percent of the box that is empty is calculated using the following ratio:

$$\frac{\text{amount of empty space}}{\text{total volume of box}} = \frac{\%}{100}$$

Assessment

1. Student product:
   - completed handout
2. Observation of student group work
3. Journal questions:
   (a) What is the reason for the empty space in the cereal box?
   (b) How do cereal boxes lead you to believe that you are getting more in the box?
   (c) Why are cereal boxes made with a small base and large front when a larger base would make them less likely to tip and fall?

Extension

- Have students design a cereal box that will hold the volume of cereal in their box with the smallest surface area and no empty space. How does it differ from the original box?

Name ______________________________ Date ______________

Cereal Box Study

Use a centimeter ruler to find the following measurements of your cereal box. Round the measurements to the nearest centimeter.

Height ____________ cm.

Width ____________ cm.

Length ____________ cm.

Use the measurements of your box to calculate the volume of your box. Volume is found by multiplying the height × width × length.

The volume of the cereal box is ______________ cubic centimeters.

Find the surface area of your box by finding the area of each side and adding the results. Area equals side times side.

Area of front = ____________ cm^2
Area of back = ____________ cm^2
Area of right side = ____________ cm^2
Area of left side = ____________ cm^2
Area of top = ____________ cm^2
Area of bottom = ____________ cm^2
The total surface area of my box is ____________ cm^2.

Use the nutrition label on your cereal box to complete the following information.

What is the size of one serving in cups? ______________

How many servings are in the box? ______________

How many cups of cereal are in the box? ______________

One cup equals 240 cubic centimeters. How many cubic centimeters of cereal are in your cereal box? ______________

Subtract the amount of cereal in the box from the volume. How many cubic centimeters of empty space are there in the box? ______________

What percent of the cereal box is empty? ______________

Burn Those Calories

Areas of Study

Reading charts and tables, computation, decimal ratios, time measurement, problem solving, decision making

Concepts

Students will:

- calculate decimal ratios
- use charts and tables to find information
- calculate personal calories burned during exercise
- design an exercise program using personal data

Materials

- Burn Those Calories handouts
- calculator
- additional resource books containing exercise and calories-burned data

Procedures

The handouts for this activity contain a table that lists the number of calories a 100-pound person would burn in one hour of aerobic exercise. Although this list is not extensive, it does contain the most common forms of exercise. You may wish to collect a more extensive list for students to use.

A 100-pound person was used in the example to make computation easier. A decimal ratio is formed during computation which does not require the use of calculators. For example, bicycling at 6 mph burns 160 calories/100 lb. or 1.6 calories/1 lb. A 135-pound person could compute calories burned during bicycling by multiplying weight (135) by the decimal ratio (1.6): $135 \times 1.6 =$ 216 calories burned each hour.

An imaginary pig-out day is the next part of the lesson. This person ate 1,118 calories above the optimal 2,000 calories. Students are asked to use their personal calorie-burning data to develop an exercise plan.

Assessment

1. Student product:
   - completion of exercise plan
2. Observation of students
3. Journal question:
   (a) Explain, besides the excess calories, why the pig-out day menu was nutritionally unsound.

Extensions

- Students can develop a personal exercise plan that includes aerobic exercise at least four times per week. Charts can be designed to help students keep track of their exercise schedules and other pertinent data.
- Students who participate in sports can calculate the number of calories burned during their practices and game days. They can then be asked to explain why, very often, retired athletes put on so much weight.

Name ______________________________ Date ______________

Burn Those Calories

You now know what you should be eating. But what if you have a pig-out day? Do you have to go on a diet? Suppose you exercise?

How many calories are burned during aerobic exercise? Everyone burns calories at a different rate. However, you can get an idea of how many calories you are burning per hour during your exercise if you take a look at the numbers below. You may want to use this data to help you develop your perfect menu—remember, exercise burns calories, so you may be able to have an extra dessert!

Directions: Use the information in the table below to find the calories you would burn per hour per pound of your weight. Set up a ratio like this:

$$\text{Calories bicycling 6 mph} = \frac{160}{100} = 1.6$$

A 143-pound person would burn 143×1.6, or about 229 calories in one hour. Find the decimal ratio for the activities below and then find the number of calories burned per hour for your weight.

Activity	100-lb person	Calories burned per 1 hr per 1 lb	Calories I would burn per hour for my weight
Bicycling, 6 mph	160	1.6	
Bicycling, 12 mph	270		
Jogging, 5.5 mph	440		
Jogging, 7 mph	610		
Jumping rope	500		
Running in place	430		
Running, 10 mph	850		
Swimming, 25 yards/min.	185		
Swimming, 50 yards/min.	325		
Tennis	265		
Walking, 2 mph	160		
Walking, 3 mph	210		
Walking, $4\frac{1}{2}$ mph	295		

These figures are taken from the American Heart Association's *Exercise Diary*. They show the approximate calories burned per hour by a 100-pound person as a result of a regular exercise plan.

Name ______________________ Date ______________

Burn Those Calories

Now here's your pig-out day! You've spent the day eating at various fast-food restaurants and have really had an off day.

Use the information on the table of exercises to plan a week of exercise to help burn the calories you ate. Be sure to write down the amount of time and the speed you will be exercising.

Meal	Foods Eaten	Calories
Breakfast at McDonald's	Orange juice	83
	Egg McMuffin	290
	Chocolate milk	210
	TOTAL	
Lunch at McDonald's	Big Mac	560
	Side salad	60
	Large french fries	400
	Cola	120
	TOTAL	
Dinner at Pizza Hut	Salad with thousand island dressing	120
Dessert at Baskin-Robbins	Personal Pan Pizza with pepperoni	675
	2 scoops Baskin-Robbins rocky road	600
	TOTAL	

You want to eat about 2,000 calories each day. How many calories did you eat on this day? ______________________

How many more calories is this than your optimal 2,000? ______________

Use the table on the next page to design an exercise program to burn up these excess calories.

Name ______________________________ Date ______________

Burn Those Calories

Now plan an exercise program for the week, using the exercise chart you filled out. Remember, you can't burn all the calories in one day, so plan on taking the whole week to do it. You can decide to exercise for only three days rather than every day. Plan ahead.

Day	**Exercise**	**Amount of Time**	**Calories Burned**
Monday			
Tuesday			
Wednesday			
Thursday			
Friday			
Saturday or Sunday			
TOTAL NUMBER OF CALORIES BURNED			

Do you think a schedule like this might help you with an exercise program? ______

Explain your answer. ______________________________

____

____

____

Cardiovascular Fitness

Areas of Study

Data collection and analysis, computation, calculation of mean, time, percent, problem solving

Concepts

Students will:

- work collaboratively to collect personal data
- find the mean for their group
- calculate heartbeats per year
- use percent to calculate target heart rate zone

Materials

- Cardiovascular Fitness handouts
- calculator
- watch with second hand
- an outdoor area or large room to exercise

Procedures

During this activity, students will be asked to find their heart rates. They should be instructed to use their wrist pulses or the carotid artery found on the side of the neck. You may want to model the correct method to make sure students do this safely. Be sure each student has found his or her pulse before continuing. Students should not be using their thumbs or they will get a double pulse.

Working in groups of four, each student counts his or her own heartbeat for 15 seconds. This count is multiplied by four to find the beats per minute. Conversions are made to beats per day and per year. Students can use 365.25 to account for leap years.

This experiment is repeated after students exercise (jog or power walk) for 3 minutes. They are asked to calculate the increase in heart rate for themselves and the average for the group.

After doing this experiment, students are asked to calculate their target heart rate. This is calculated using a specific formula explained on the Target Heart Rate page. For most people, the target is calculated at 70% of the difference between 220 and the person's age. For those in excellent physical condition, it is calculated at 85%.

Possible Solution

Assume this target heart rate is being calculated for a 20-year-old.

$220 - 20 = 200$

$200 \times .70 = 140$

$200 \times .85 = 170$

During aerobic exercise, this person's heart rate should be between 140 and 170 beats per minute or between 35 and 43 beats every 15 seconds.

Assessment

1. Student products:
   - data sheets and target heart rate
2. Observation of students: Observe and question groups at work
3. Journal questions:
   - Why do you think people have different heart rates? Can you think of ways to explain the differences that occurred in your group?

Extension

- Lesson can be extended to include cooldown time. After exercising, students observe how long it takes them to return to their at-rest heart rates.

Name ______________________________ Date ______________

Cardiovascular Fitness

Imagine a pump that weighs about a half pound (about the size of a fist). This pump works 24 hours a day for about 70 years. It has the power of a champion athlete. Every hour it produces enough energy to raise a small car off the ground. This amazing pump is the human heart. The average healthy adult heart beats about 72 times per minute. Do you think your heart beats faster or slower than the average? How many times do you think your heart beats in one day? In one year? Let's do an experiment to find out.

Directions: Find the pulse rate of each member of your group. Use the pulse at your wrist or the carotid artery on the side of your neck. Remember to press lightly on your neck. Be sure you do not use your thumb or you will get a double pulse. Use a watch with a second hand and count the number of times your heart beats in 15 seconds. Use this data to figure out the beats per minute, per day, and per year.

Resting Heart Rate				
Your Name	**Beats per 15 seconds**	**Beats per minute**	**Beats per day**	**Beats per year**
Average				

Use the average from the table above to figure out how many times the average heart would beat in 80 years. Explain how you solved this problem. (Use the reverse side of this sheet for your answer.)

Name ______________________________ Date ______________________

Cardiovascular Fitness

While the average heart beats 72 times per minute, the normal range is between 60 and 100 beats per minute. However, physical exercise can make your heart beat faster. What do you think will happen to **your** heart rate if you exercise? Let's find out.

Directions: Work with your group again to collect this data. You will need to exercise for 3 minutes. You can jog or power-walk. If you are power-walking, be sure to really walk! Write down your new data on the table below. Use this information to find the increase in your heart rate.

Jogging or Power-Walking Heart Rate				
Your Name	**Beats per 15 seconds**	**Beats per minute**	**Increase in beats per min.**	**% of increase**
Average				

How did the average pulse rate change after this exercise? ______________

Explain your answer here. ______________________________

____

____

Name ______________________________ Date ________________

Cardiovascular Fitness

Target Heart Rate

You can do a simple mathematical test to determine if an exercise is aerobic (helping to strengthen your heart and lungs). In this activity, you will compare your exercise heart rate from before with the target heart rate zone.

Steps to finding your target heart rate zone:

Step 1. 220 – your age = ________________

Step 2. Multiply the result of step 1 by 70% (.70) ________________

Step 3. Multiply the result of step 1 by 85%. (.85) ________________

My target heart rate zone is between ______________________ and
(answer to Step 2)

______________________.
(answer to Step 3)

Was your rate after exercise in your target heart rate zone? ________________

Worldwide Nutritional Concerns

Areas of Study

Data collection, computation, problem solving, research skills

Concepts

Students will:

- design a 3,500-calorie-a-day diet and a 1,000-calorie-a-day diet after researching a developing country
- use charts and tables to find information
- calculate total calories and grams of protein in one day's diet

Materials

- Worldwide Nutritional Concerns handouts
- calculator
- additional resource books containing information about poor countries

Procedures

Read through and discuss the Worldwide Nutritional Concerns and Research Project with students. It is suggested that students work in pairs to facilitate the research and split up the tasks into more manageable pieces.

Students will need to research the customs and traditions of their countries before they can develop a menu. For example, if beef is not eaten by the majority of the people in a country, then this cannot be part of the diet. In addition, not all foods that are available in the United States are available worldwide. Religious considerations and locally available food supplies should also be factored.

This type of research can be extensive, and you may want to integrate it with social studies and language arts, if possible.

Assessment

Assessment will be based on the completeness and accuracy of the research data and the accurate interpretation of nutritional values.

1. Student product:
   - The research project should be a major segment of the assessment.
2. Observation of students: Was the work evenly distributed among team members?
3. Journal question:
   - Explain your reactions to the two menus developed. What were the most dramatic differences between the two?

Name ______________________________ Date ______________

Worldwide Nutritional Concerns

We have learned that our bodies need a certain amount of calories. We also need proteins, carbohydrates, and fats. Much about what you eat and how you think about food depends on where you live.

In Europe and the United States, people eat about 3,500 calories each day. They also eat about 20 times more sugar and 5 times more fat than they did 100 years ago.

In developing areas, such as some countries in Africa and Asia, people eat an average of less than 1,000 calories per day. The diets in these areas are made up mostly of beans, vegetables, and grains. These diets may be very low in protein, vitamins, and minerals. In the United States, many people are overweight because each person, on average, eats 1,250 **more** than the recommended amount of calories each day. In third-world countries, there is a problem with malnutrition because, on average, each person eats 1,250 calories **less** than the recommended amount. What might a typical diet in a poor country look like?

Directions: Use the table of calories or reference books to develop a menu for two people, one living in the United States, and the other living in a poor nation. You will need to list only the calories and the grams of protein each person eats. Be sure to research the kind of diet a person in the poor country you choose might really have.

USA			**Developing Country**		
Food	**Calories**	**Protein (g)**	**Food**	**Calories**	**Protein (g)**
TOTAL	at least 3,500		TOTAL	at least 1,000	

Name ______________________________ Date ______________

Worldwide Nutritional Concerns

Research Project

Names of Partners ______________________________

____

The country we are researching is: ______________________________

Reference books or sources we plan to use: ______________________________

____

____

____

Work with your partner to compare a day in the life of a person in the United States with that of a person from the country you have chosen. Think about these questions to help you get started:

1. How might a person in the country you have chosen spend his or her day?
2. How might they think about food?
3. How much money does the average person in that country make?
4. Where do the people in that country go to get their food?
5. Are there any foods that are not allowed by some religions there?
6. How long do people in that country usually live?
7. Are there times when food is not available there? Why?
8. How does the diet in that country affect the health of the people there?
9. What can I do to help alleviate world hunger?

Use reference books to learn about the culture of the country you have chosen. This will help you talk about the life of one person who lives there. You and your partner must give a complete description of two people—one from the United States and one from a third-world nation. Be sure to do a great deal of research before you begin.

The Distribution of Food

Areas of Study

Data collection and analysis, reading tables and charts, computation, percents, problem solving

Concepts

Students will:

- calculate percents
- use charts and tables to find information
- participate in a simulation
- write about their experiences

Materials

- The Distribution of Food handouts
- calculator
- 1 cookie (Oreo® cookies work well)

Procedures

Read through the information on the population table with students. Work through the first problem with them to make sure they all understand how to find the percent of population. To find the percent of population for Africa:

$$\% \text{ of population} = \frac{\text{population of Africa}}{\text{total world population}}$$

$$\% \text{ of population} = \frac{732}{5{,}451} \approx 0.13 = 13\%$$

Have students complete the table and calculate the combined percentages for the countries. The activity develops in the following way:

- The assumption is made that one cookie represents the required amount of food for a healthy diet. Any more would result in obesity, and any less would result in malnutrition and possible starvation.
- Each student is given one cookie but told not to eat it because it will be redistributed based upon the actual distribution of food.
- 85% of the world's population (85% of the number of students in your class) need to be given 40% of the available food. If there are 20 students in class, 17 will be from Africa, Asia, and Latin America. They will have only 8 cookies (40% of the world's supply).
- 15% of the world's population (15% of the number of students in your class) will be given 60% of the available food. Three students are in this group, and they have 9 cookies.
- This will mean that the students who represent Europe and North America will have too much food and will have problems with obesity. Students who represent Africa, Asia, and Latin America will not have enough cookies. Depending on their method of distribution, all will be hungry but some may be the victims of starvation.

If you can conduct the simulation with 100 students, it is much more effective. Each cookie represents 1% of the food supply and each student represents 1% of the world's population.

Assessment

1. Student products:
   - worksheet and written statement
2. Observation of students
3. Journal question:
   - Explain how you calculated the number of cookies each of the countries would receive.

Name ______________________ Date ______________

The Distribution of Food

Use the information below to find what percentage of the world's population lives in North America and Europe. Then find what percentage of the world's population lives in Africa, Asia, and Latin America.

Region	Population	% of World Population
Africa	732,000,000	
Asia	3,428,000,000	
Europe	507,000,000	
North America	295,000,000	
Latin America	489,000,000	
TOTAL POPULATION	5,451,000,000	

Information obtained from the 1997 *World Almanac and Book of Facts*

1. About what percentage of the world's population lives in Africa, Asia, and Latin America combined? ______________________

2. About what percentage of the world's population lives in Europe and North America combined? ______________________

Name ____________________ Date ____________

The Distribution of Food

The Simulation

You have learned that about 85% of the world's population lives in Africa, Asia, and Latin America. How much of the world's food supply do these countries receive?

- Africa, Asia, and Latin America receive only about 40% of the world's food.

About 15% of the world's population lives in Europe and North America. How much of the world's food do these countries receive?

- Europe and North America have access to about 60% of the world's food.

The experiment you are going to do now is called a simulation. This is because it simulates real-world experiences. Follow these steps:

Step 1. Count the total number of students in your class. Of this total, 85% will represent the population of Africa, Asia, and Latin America. They should go to one side of the room.

Step 2. The other 15% of the class represents the population of Europe and North America. They should go to the other side of the room.

Step 3. To begin the simulation, each student will get one cookie—this cookie represents all the food they will need for a healthy, normal life. DON'T EAT THE COOKIE BECAUSE YOU MIGHT NOT GET TO KEEP IT!

Step 4. The total food in the world is represented by all the cookies in the simulation. The cookies must be redistributed in the same percentages that food is available in the world. The people of Africa, Asia, and Latin America will get 40% of the cookies. The people of Europe and North America will get 60% of the cookies.

Redistribute the cookies now.

Questions:

1. How many cookies does each person in Africa, Asia, and Latin America get to eat each day? ____________
2. How many cookies does each person in Europe and North America get to eat each day? ____________

Name ______________________________ Date ______________________

The Distribution of Food

If you were part of the Africa, Asia, and Latin America group in the simulation, describe how you felt when you had to give your cookie to other countries.

If you were one of the lucky ones who lived in Europe or North America, describe how you felt when you took the cookies from the other people.

A Nutrition Word Search

Procedure

Have students search for the following nutritional words.

BREAD
CALORIES
CANDY
CARBOHYDRATES
CEREAL
CHOLESTEROL
EGGS
EXERCISE
FAT
FIBER
FISH
FRUIT
GRAINS
HEALTHY
LOW FAT
MEAT
MILK
MINERALS
NUTRITION
PASTA
POULTRY
PROTEIN
SALAD
SATURATED
SERVING
SODIUM
SUGAR
VEGETABLE
VITAMINS

Solution

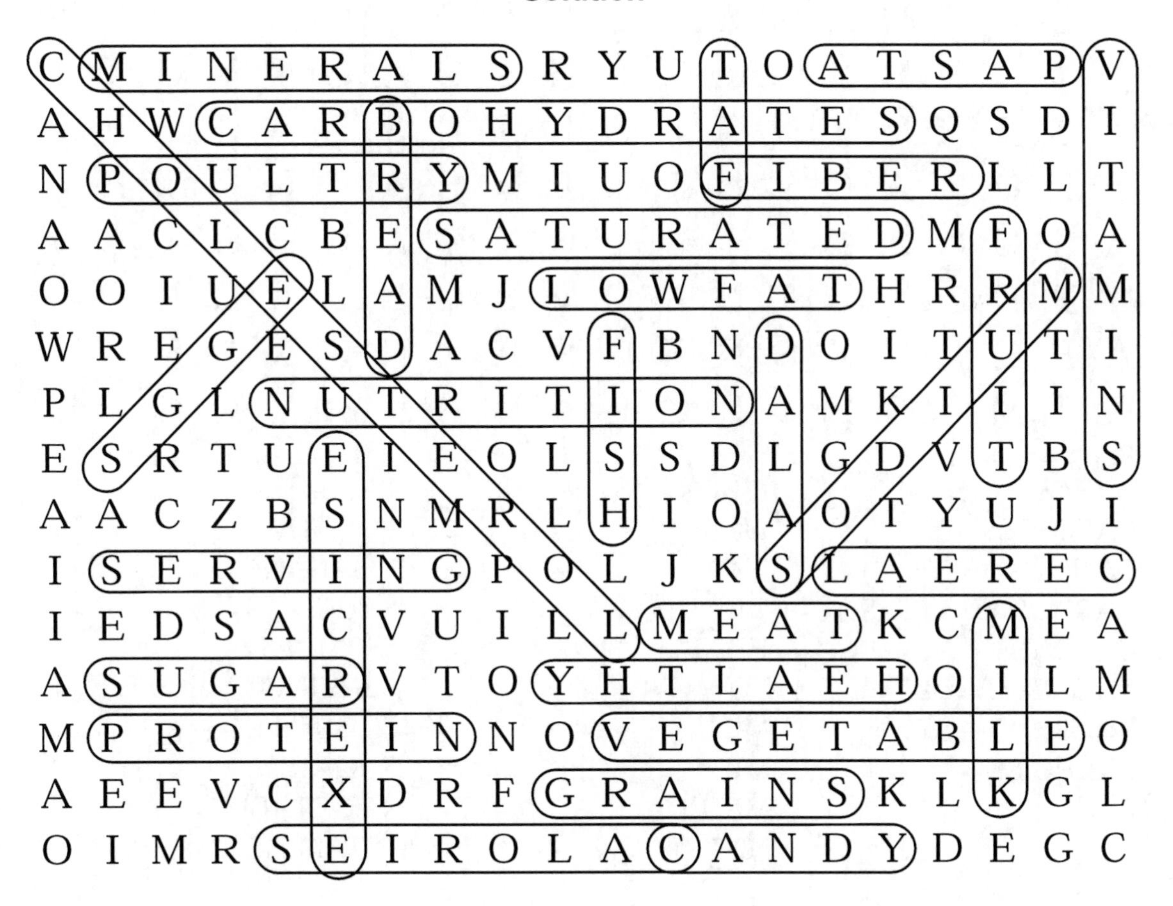

A Nutrition Word Search

Search this rectangle of letters for the words listed below. Circle the words when you find them.

C	M	I	N	E	R	A	L	S	R	Y	U	T	O	A	T	S	A	P	V
A	H	W	C	A	R	B	O	H	Y	D	R	A	T	E	S	Q	S	D	I
N	P	O	U	L	T	R	Y	M	I	U	O	F	I	B	E	R	L	L	T
A	A	C	L	C	B	E	S	A	T	U	R	A	T	E	D	M	F	O	A
O	O	I	U	E	L	A	M	J	L	O	W	F	A	T	H	R	R	M	M
W	R	E	G	E	S	D	A	C	V	F	B	N	D	O	I	T	U	T	I
P	L	G	L	N	U	T	R	I	T	I	O	N	A	M	K	I	I	I	N
E	S	R	T	U	E	I	E	O	L	S	S	D	L	G	D	V	T	B	S
A	A	C	Z	B	S	N	M	R	L	H	I	O	A	O	T	Y	U	J	I
I	S	E	R	V	I	N	G	P	O	L	J	K	S	L	A	E	R	E	C
I	E	D	S	A	C	V	U	I	L	L	M	E	A	T	K	C	M	E	A
A	S	U	G	A	R	V	T	O	Y	H	T	L	A	E	H	O	I	L	M
M	P	R	O	T	E	I	N	N	O	V	E	G	E	T	A	B	L	E	O
A	E	E	V	C	X	D	R	F	G	R	A	I	N	S	K	L	K	G	L
O	I	M	R	S	E	I	R	O	L	A	C	A	N	D	Y	D	E	G	C

BREAD
CALORIES
CANDY
CARBOHYDRATES
CEREAL
CHOLESTEROL
EGGS
EXERCISE
FAT

FIBER
FISH
FRUIT
GRAINS
HEALTHY
LOW FAT
MEAT
MILK
MINERALS
NUTRITION

PASTA
POULTRY
PROTEIN
SALAD
SATURATED
SERVING
SODIUM
SUGAR
VEGETABLE
VITAMINS

Nutrition Crossword

Procedure

Have students use the nutritional clues provided to fill in the crossword puzzle.

Solution

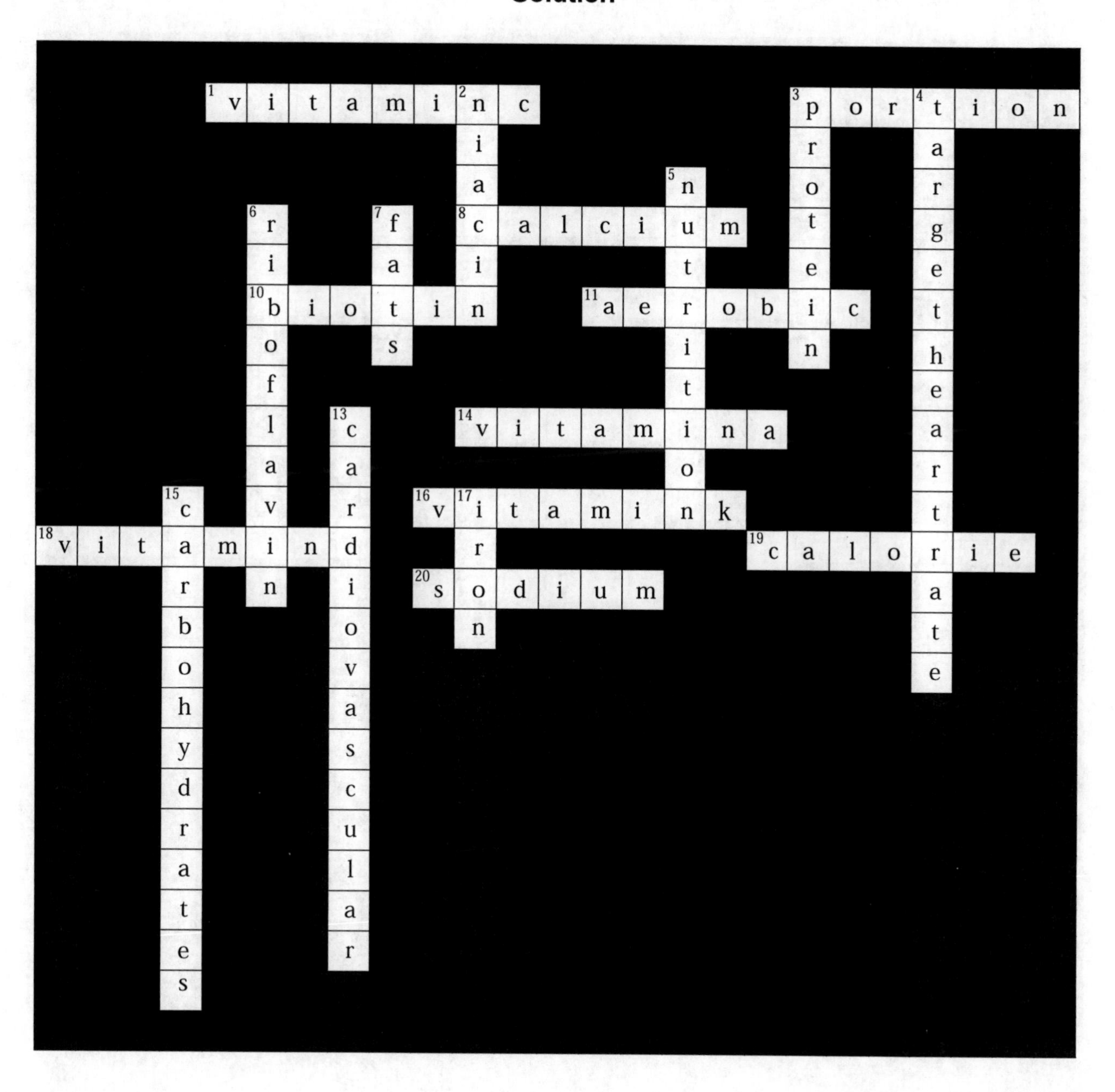

Name ______________________________ Date ______________________

Nutrition Crossword

Use the new words you have learned about nutrition to do this crossword puzzle. The clues on the following page will help. If the answer is two words, no space has been left between them. For example, vitamin A would be written as vitamina.

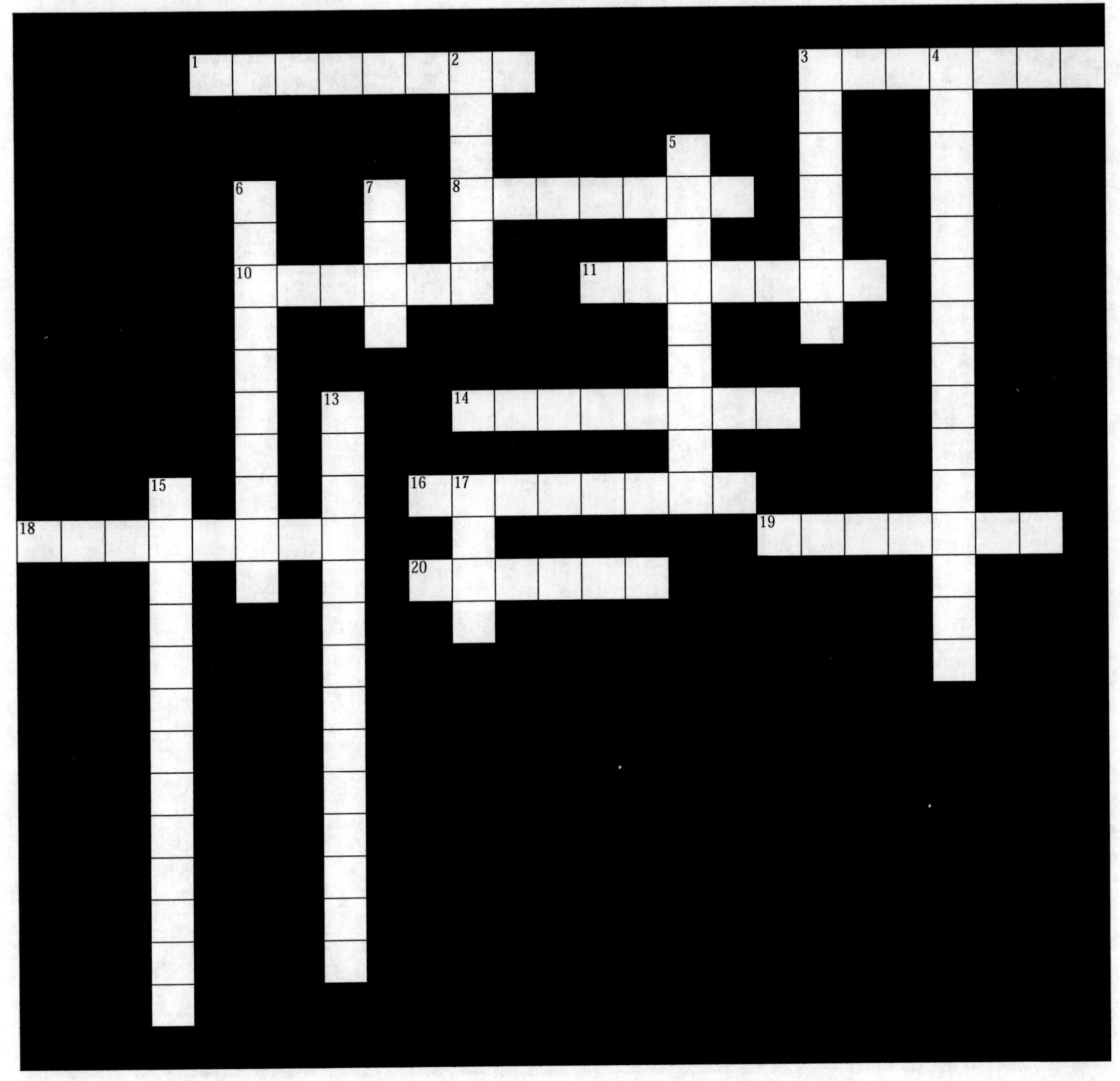

Name ______________________________ Date ______________

Nutrition Crossword

Across

1. This ascorbic acid is found in citrus fruits.
3. This is a single helping of food.
8. This mineral found in dairy products helps build strong bones.
10. Vitamin H.
11. This is an activity that burns calories and causes the heart to beat faster.
14. This nutrient is found in fish-liver oil and some green and leafy vegetables.
16. This nutrient is found in leafy green vegetables, tomatoes, and vegetable oils. It is essential for blood clotting.
18. This nutrient is made by the body from exposure to sunlight. Lack of it causes rickets in children.
19. This is a measure of the amount of energy found in food.
20. This is an element found in salt. Its chemical symbol is Na.

Down

2. This is another name for nicotinic acid. It is a member of the vitamin B complex.
3. This nutrient is found in meat, poultry, fish, eggs, and nuts. You should eat 2 to 3 servings of it per day.
4. This is the number of heartbeats per minute that is the goal of an exercise program (3 words).
5. This is the name for the study of foods people eat and need.
6. This is part of the vitamin B complex, also known as vitamin B_2.
7. These are oily compounds found in plants and animals.
13. This word is used to talk about the heart or blood vessels.
15. This group of nutrients includes sugars, starches, and cellulose. They are made up of carbon, hydrogen, and oxygen.
17. This mineral helps the red blood cells transport oxygen.

Nutrition and Poetry

Procedure

One way to integrate mathematics with other subject areas is to include some interesting poetry assignments as part of the math lesson. The four poems included in Nutrition and Poetry are acrostic, cinquain, haiku, and form.

Acrostic poems have the letters of a word written vertically; each line of the poem begins with the corresponding letter of the title. A word, a phrase, or a sentence can be written on the line. The line must refer to the title of the poem and cannot merely be a collection of words that begins with that letter.

Cinquain poems consist of five lines. The first line is a one-word line, the second line is a two-word line, the third line is a three-word line, the fourth line is a personal reaction of the author to the topic, and the fifth line is a single-word synonym of the first line. Each of the words must refer to the title (the first line).

Haiku is a three-line poem in which each line has a definite syllable count. Line one has 5 syllables, line two has 7 syllables, and line three has 5 syllables.

Form poems take the shape of the object of the poem, and the words relate to the subject.

An example of each type of poem is included in the lesson, and students are asked to compose a poem of their own.

Assessment

The following matrix is suggested as a means of assessing the poetry:

Criteria	4	3	2	1
How well did the student follow the guidelines for the poetry type?				
How correct is the grammar?				
How creative is the poem?				
What is the overall quality of the project?				

Name ______________________________ Date ______________

Nutrition and Poetry

Ode to Vegetables: Acrostic

Select a vegetable, fruit, or other nutritional term as the title of the poem. Write the letters of the title vertically. Each line of the poem begins with the corresponding letter of the title. A word, phrase, or sentence can be written on the line. Each line of the poem must refer to the meaning of the title.

Example:

Carrot
A vitamin source
Rabbit's favorite
Raw, boiled, or
Oven-roasted, they become a
Tasty, nutritious treat.

Your acrostic:

Name ______________________________ Date ____________________

Nutrition and Poetry

Ode to Vegetables: Cinquain

This poem is made up of five lines. The first line is a single vegetable, fruit, or nutritional term naming the topic of the poem. The second line has two words that describe the first line. The third line has three words that give more detail about the first line. The fourth line is a personal reaction of the author to the topic. It tells how the author feels about the subject. The fifth line is a single-word synonym for the first line.

Example:

Potato
Root crop
Fried, mashed, scalloped
Baked is my favorite
Spud

Your Cinquain:

Name ______________________________ Date ______________

Nutrition and Poetry

Ode to Vegetables: Haiku

Haiku poems have a very definite syllable count. These poems had their origin in Japan. This is the structure of a haiku:

Line one:	5 syllables
Line two:	7 syllables
Line three:	5 syllables

Example:

Zucchini

We love zucchini
Tasty squash upon a vine
Can't give them away!

Your Haiku:

Name ______________________ Date ______________

Nutrition and Poetry

Ode to Vegetables: Form Poem

Form poems are designed to take the shape of the subject of the poem. Our example of a form poem is in the shape of a beet, because the subject is beets. Your poem must have a fruit, vegetable, or other nutritional subject as its topic and its shape!

Ode to Beets

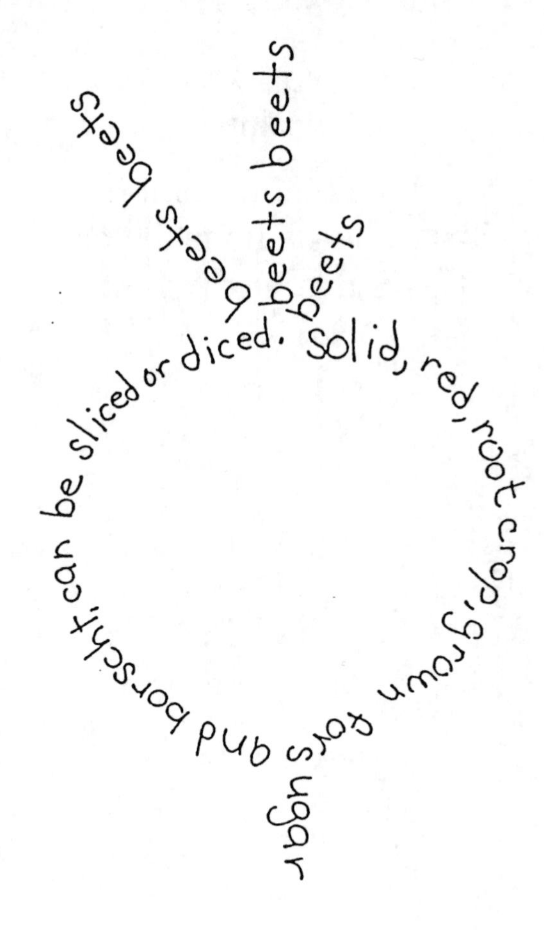

Name ______________________________ Date ______________

Nutrition and Poetry

Ode to Vegetables

Design your own form poem. Construct the finished poem in the space below.

Name ____________________ Date ____________________

Nutrition and Music

The song below, "Protein," is a little ditty for your enjoyment. While you sing it, keep in mind what it has to say about the body's building blocks. If you want to sing along, the words are on the next page.

Name ______________________________ Date ______________

Nutrition and Music

Protein

Meat and fish are some,
Cheese and milk are too.
Can you guess what we are?
Protein, protein,
P-R-O-T-E-I-N.
We make your bones strong,
We build up tis-sue,
You need us, yes you do.
Protein, protein,
P-R-O-T-E-I-N.

Nutrition and Art

Procedure

Place a transparency of this diagram of the digestive system on an overhead projector. Have students copy the diagram on an $8\frac{1}{2}" \times 11"$ sheet of paper. They can then use Crayola® Fabric Crayons to color in the diagram, following the directions on the crayons to transfer the design, to make their very own T-shirts.

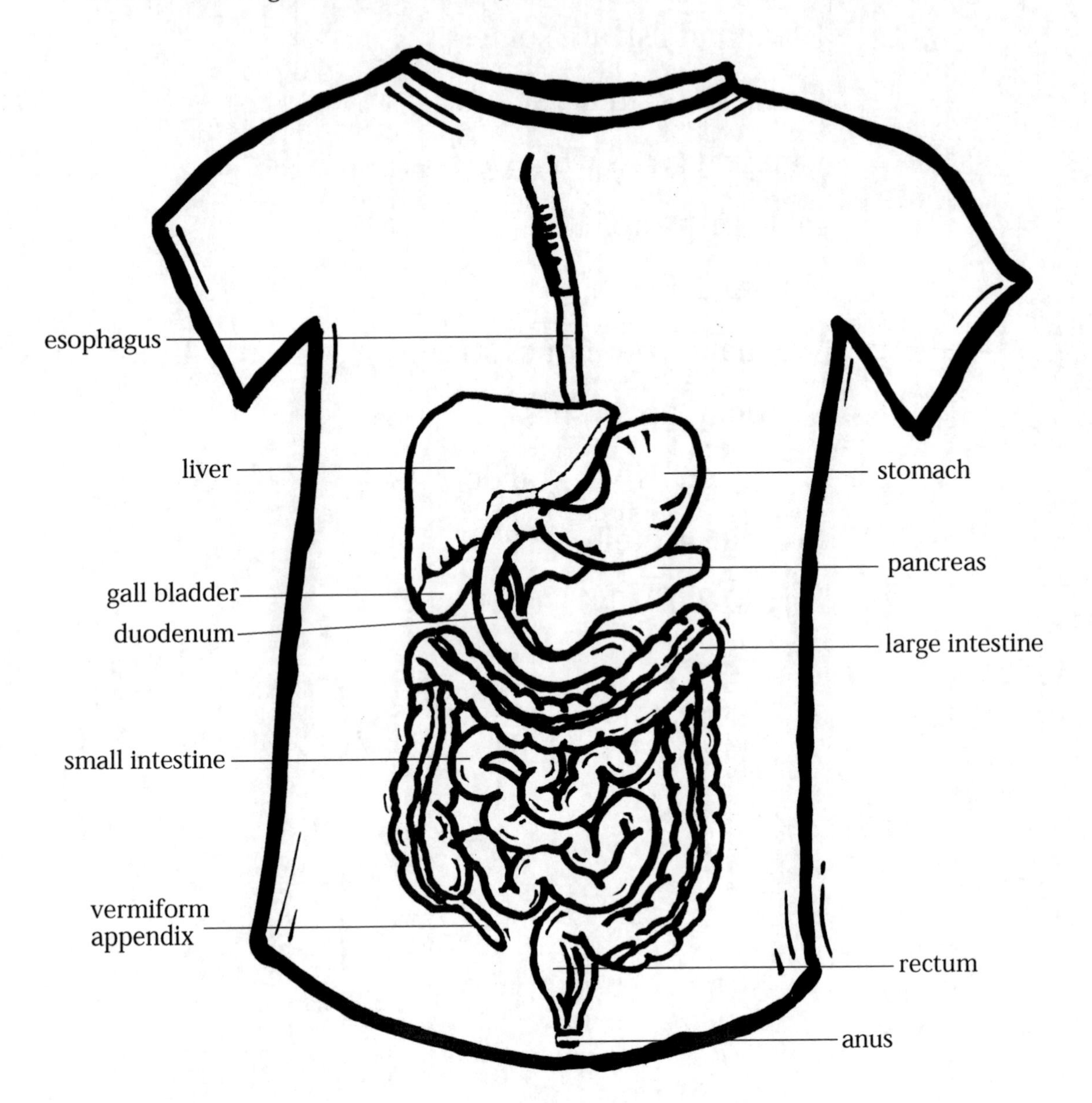

Research Projects

Procedure

There are many projects to help students (1) improve their research skills, (2) answer questions about the nutritional value of foods, and (3) see how the world of marketing has influenced and affected their diet and health. The projects described in this section are suggestions of motivating projects that allow students to make individual choices.

Analysis of Snack Foods requires students to identify 25 snack foods and label them as "healthy" or "unhealthy." Which of these foods are more expensive? Which would be considered a "good buy"?

Dinner Label Project asks students to plan a dinner menu and to make a collection of the nutrition labels for the primary or main ingredients. If they ate this meal, would they be fulfilling their daily requirements?

Historical Nutritional Concerns asks students to explore a food-supply catastrophe from the past and analyze the cause, cure (if any), and possible prevention. This project asks students to become an historical nutritionist and utilize their critical thinking skills.

A Pyramid Cookbook is a creative project in which students design a cookbook having the number of recipes in each category reflect the number of servings one should eat as defined on the Food Pyramid. By reproducing the book, the student's work is validated and excellence encouraged.

An Advertising Campaign will appeal to students who possess creativity and an interest in marketing. Most advertising promotes sugars, fats, and, in general, unhealthy foods. Students are asked to design an advertising campaign for a healthy food, such as broccoli.

A Consumer Research Project is an interesting way to have students critically examine the choices we are given when we purchase a particular food item. We can buy a brand-name peanut butter, a store-brand, or a generic brand. What are the relative costs? Is there a difference in quality? What is the "best buy" and what does this mean? Is the best buy always the cheapest one?

Supermarket Design encourages students to look at the design or layout of the supermarket, using a critical eye. What foods are at the front of the store? What foods are at eye level? What items are placed right next to the cash registers and catch our eye when we are waiting to check out? If students had a product to sell, where would they want it displayed? This project encourages critical thinking.

Name ______________________________ Date ______________________

Research Projects

1. **Analysis of Snack Foods**
   Research the cost per pound of 25 different snack foods. Include both healthy snacks, such as fruit, and not-too-healthy snacks, such as candy and chips. Here is an example of a table showing data for three different snack foods. The healthy snack is checked.

Snack Food	Healthy?	Weight	Cost/Unit	Cost/Pound
Jelly beans		14 oz	$1.59	$1.82
Tortilla chips		14.5 oz	$1.29	$1.42
Apple	✓	4.5 oz	$0.18	$0.79

   Make a table containing 25 different snack foods. Then discuss the nutritional value and cost of the snacks you listed.

2. **Dinner Label Project**
   You will be planning a dinner menu. Write down the menu and list all the ingredients you will need. You may need to check recipes to get all of them. Now collect all of the nutrition labels from each food. The labels will have a great deal of information you can use to answer this question: What percent of your daily nutritional requirements would you get from your menu?

3. **Historical Nutritional Concerns**
   Throughout history there have been many famines. Also, because of the lack of knowledge of what vitamins and minerals are needed for healthy development, many diseases have occurred that could have been prevented with good nutrition.

   Research a nutritional problem from the past. This could involve a famine or a disease and its cure or solution. Who discovered the solution? What methods were used to prevent a recurrence? With this report you will become a historical nutritionist.

Name ____________________ Date ____________

Research Projects

4. **A Pyramid Cookbook**
   Design a cookbook (either ethnic, regional, or vegetarian) that reflects the suggested servings from the Food Pyramid. That means that you will need:

   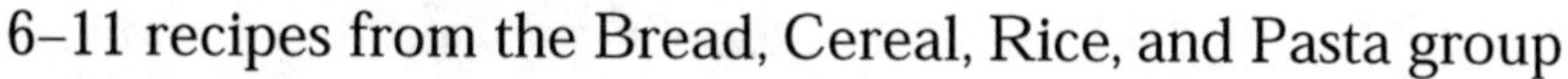

   - 6–11 recipes from the Bread, Cereal, Rice, and Pasta group
   - 3–5 recipes from the Vegetable group
   - 3–5 recipes from the Fruit group
   - 2–3 recipes from the Milk, Yogurt, and Cheese group
   - 2–3 recipes from the Meat, Poultry, Fish, Eggs, and Nuts group
   - 1–2 recipes from the Fats, Oils, and Sweets group

   Include in your book good recipes that can be used by other people in the class.

5. **An Advertising Campaign**
   Most advertisements are for foods with high amounts of fats, oils, and/or sugars. Design an advertisement for a healthy food. Discuss the benefits of the food. Use pictures to make the food look good to the consumer. Write a poem or jingle to help sell your product. Be prepared to present your advertisement to the class.

6. **A Consumer Research Project**
   Many foods come in two different packages at the supermarket: the national brand (made by a major company and packaged by them) and a store brand (packaged and marketed by the supermarket). Choose one food item (such as peanut butter, orange juice, or canned vegetables) that comes in a national brand and a supermarket brand. Compare the nutritional values, the price per unit, the appearance, and the taste of each brand.

   To help with taste, conduct a blind taste test using at least 10 people. Make sure they do not know which of the brands they are tasting. This is a blind taste test. Is one brand more popular than the other? Use a table to organize your data and a graph to display your results. Write an analysis of the results. Be prepared to present your data to the class.

7. **Supermarket Design**
   Take a field trip to your local supermarket. Make a diagram of the store layout. Show where different foods can be found—what is near the front of the store, what is displayed at the ends of the aisles, what is on the top shelf, what foods are found right at eye level, and what is located at the cash register. How are consumers encouraged to buy more? Place your diagram on a poster board and, remember, neatness counts.

Web Sites, Books, and Pamphlets

Books and Articles

American Association for the Advancement of Science. (1989). *Project 2061: Science for All Americans.* Washington, DC: AAAS. Unites mathematics, science, and technology in a definition of scientific literacy. The report focuses on four areas: (1) achieving reform in these three subjects, (2) improvement of teaching, (3) professional development of teachers, and (4) restructuring schools to encourage new methods of instruction.

Fogarty, R. (1994). "Thinking about themes: Hundreds of themes." *Middle School Journal.* March, p 30-31. Discusses how and why to develop themes and integrate them into the middle school program.

National Council of Teachers of Mathematics. (1989). *Curriculum and Evaluation Standards for School Mathematics.* Reston, VA: NCTM. Delineates a change in the manner in which mathematics is taught and assessed in our changing world.

Natow, A., and Heslin, J. (1985). *No-Nonsense Nutrition for Kids.* New York: McGraw-Hill & Co. Written in question-and-answer format. This book is divided into three sections: Part 1 talks about the needs of children at various stages of development, Part 2 relates special situations, and Part 3 contains recipes.

Natow, A., and Heslin, J. (1995). *The Supermarket Nutrition Counter.* New York: Pocket Books. Contains the calorie, fat, sodium, carbohydrate, and fiber values for more than 16,000 items you would find on supermarket shelves. If you want to know the nutritional value of Mrs. Paul's Crunchy Fish Sticks, this book has the answer.

Netzer, C. T. (1992). *Encyclopedia of Food Values.* New York: Dell Books. A comprehensive reference that includes:

- a huge listing of fresh, frozen, prepared, brand-name, and fast foods
- listings for 16 essential vitamins and minerals
- updated dietary recommendations
- calories, protein grams, carbohydrate grams, fat grams, cholesterol, sodium, and fiber for each listed food

Null, G. (1983). *Gary Null's Nutrition Sourcebook for the 80's.* New York: Collier Books. Contains tables and charts describing food composition and nutrition.

Pollak, H. (1987). Speech given at Mathematical Sciences Education Board Frameworks Conference as reported in *Curriculum and Evaluation Standards for School Mathematics* (p. 4).

Szilard, P. (1987). *Food and Nutrition Information Guide.* Littleton, CO: Libraries Unlimited. Includes bibliographies and indexes to nutrition and nutrition information services, as well as diet and food information services. An extensive information reference.

Fiction and Musical Experiences

Burstein, J. (Cassette & Book). (1989). *Slim Goodbody's Nutrition Edition.* Long Branch, NJ: Kimbo Educational. Charming vocals and music to 13 songs: On the Farm; The B-R-E-A-K-F-A-S-T High; Chomp, Chomp, Chew, Chew, Chew; Workers in the Body; The Journey of Food; Eat a Rainbow; Something Green; Your Diet, Don't Deny It; Protein Power; The Vitamin ABC's; Magical Minerals; Water Works; and Vegetable Stew.

DeGroat, D. (1992). *Annie Pitts, Artichoke.* New York: Simon & Schuster Books for Young Readers. Annie Pitts and her sworn enemy, Mathew, get into trouble on a field trip to the grocery store. This makes Mrs. Goshengepfeffer, her teacher, very angry, and Annie is assigned the least appetizing role in the school nutrition play.

Leedy, L. (1994). *The Edible Pyramid.* New York: Holiday House. On the opening day of the grand opening of the restaurant "The Edible Pyramid," customers are lined up to get inside. Customers are led through the pyramid and become expert in choosing the correct foods in just the right quantities.

Web Sites

http://www.agric.org
An extensive 49-page list of books, lesson plans, videos, games, posters, and pamphlets available for loan. A brief summary of each resource is given.

http://www.amhrt.org
The American Heart Association's Web pages contain recipes, healthy diets, exercise tips, healthy heart articles, and updates on the latest research.

http://research.med.umkc.edu/aafp/pQ.html
The Healthy Education Program lists sources of information that have been approved by the American Academy of the Family Physicians Foundation. Extensive index and sources.

http://www.healthy.net
Information about fitness, exercise, and diet. Many articles by different doctors about nutrition and fitness.

http://arborcom.com/
Very complete index and links to other nutrition sites.

http://ericir.syr.edu/Projects/Newton/10/lessons/DietNut.html
Nutrition lessons designed by the television program *Newton's Apple* for students.

http://www.ces.msstate.edu/pubs/pub1908.htm
A complete guide to understanding food labels.

http://www.nal.usda.gov/fnic/Dietary/9dietgui.htm
A complete list of the recommended daily requirements and food sources.